Thais Santos
Thayza Paiva
Gabriele Grimm

Audiovisual Production

Thais Santos
Thayza Paiva
Gabriele Grimm

Audiovisual Production

Exercises for Parkinson's patients

ScienciaScripts

Imprint

Any brand names and product names mentioned in this book are subject to trademark, brand or patent protection and are trademarks or registered trademarks of their respective holders. The use of brand names, product names, common names, trade names, product descriptions etc. even without a particular marking in this work is in no way to be construed to mean that such names may be regarded as unrestricted in respect of trademark and brand protection legislation and could thus be used by anyone.

Cover image: www.ingimage.com

This book is a translation from the original published under ISBN 978-613-9-61489-9.

Publisher:
Sciencia Scripts
is a trademark of
Dodo Books Indian Ocean Ltd. and OmniScriptum S.R.L publishing group

120 High Road, East Finchley, London, N2 9ED, United Kingdom
Str. Armeneasca 28/1, office 1, Chisinau MD-2012, Republic of Moldova, Europe
Printed at: see last page
ISBN: 978-620-7-65709-4

Table of contents:

POSITIVE UNIVERSITY

AUDIOVISUAL PRODUCTION: Series of videos with exercises for Parkinson's sufferers using visual design.

CURITIBA

2016

DEDICATORY

We dedicate this project to God, because we owe him all our support and strength in times of great difficulty.

ACKNOWLEDGMENTS

First of all, we would like to thank our families for all their support and confidence in our potential. To our friends and fellow students for their understanding and help. To our teachers, especially Professor Gabrielle Hartmann Grimm who patiently guided us and gave shape to our ideas, to Professor Wagner Regis who, in addition to the knowledge he provided to define the design specialty that would be used in the project, gave us support and collaborated at the time of greatest need, to Professor Ana Paula Franga who awakened in us the desire to participate in the social side of Design. We are immensely grateful to the Paraná Association of Parkinson's Disease Patients for their partnership, especially Vanessa who provided us with contact with physiotherapist Patricia Nazari, whom we thank for all her affection for the project, support and participation. To the APPP patients, who enthusiastically took part in the initial research and the validation at the end of the project. Physiotherapist Daiane, from the physiotherapy department at Positivo University, for her help in finding articles on the disease. Karen Munhe, who helped us with technical knowledge during the production phase.

Thank you very much, without you this project would not have been possible!

EPIGRAPH

"When people care about each other, they always find a way to make things work out." Nicholas Sparks.

SUMMARY

According to data from the World Health Organization (WHO), around 1% of the world's population over the age of 65 is affected by Parkinson's disease. Research carried out for this project showed that the disease does not only affect the elderly, but that more and more young people are finding out that they have it. It was also discovered that younger patients look for information about the disease on the internet, while older patients turn to more old-fashioned media such as DVDs. It could be analyzed that there is no variety of material developed for this audience, and it was noticed that the material on social networks is not adequate, as it is old, "homemade" and contains information that is not in line with the patients' reality.

This project consists of producing a series of videos to reinforce the physiotherapy applied in clinics and associations for Parkinson's patients. This production aims to collaborate in the treatment with exercises, in order to promote patient autonomy in a credible material and appropriate language. The exercises do not replace physiotherapy sessions, but complement them. In order to provide credibility, the series relies on the presence of a physiotherapy professional, who participates in both the production and pre-production stages, supervising and cooperating with the choice of exercises to be reproduced and editing the script. To provide quality, the knowledge acquired during the research was used, and the planning of information aspects based on the knowledge acquired in the Design study. Based on research into the target audience, it was found that the information needed to be clear and well reinforced, as the audience has various visual difficulties. Therefore, in the production there is verbal, auditory and visual reinforcement, in addition to the representation made by the physiotherapist, we opted for pictorial reinforcements, arrows that reinforce the movements of each exercise. A readability study was carried out to choose the fonts used throughout the video, and the interventions follow the principles used in choosing the font. The project's brand (EXPA) was designed to translate what the channel is about, so in simple terms it stands for Parkinson's exercises. As concepts, autonomy and relaxation were defined, and to aid relaxation the color palette chosen works with warm and cool colors, which will be used in all the information reinforcements throughout the videos. Also based on research into the target audience, we noticed the need to produce for two different audiences: people with access to the internet will access the series via Youtube, while another part of the audience does not have this access but uses DVD media, so in order to cater for both audiences, the two types of media were produced.

We believe in the power of design to help solve society's problems, problems that go beyond profit. This project was developed to show that the social aspect of design should not be forgotten.

Chapter 1

1 INTRODUCTION

According to data from the World Health Organization (WHO), around 1% of the world's population over the age of 65 is affected by Parkinson's disease. Research carried out for this project showed that the disease does not only affect the elderly, but that more and more young people are finding out that they have it. It was also discovered that younger patients look for information about the disease on the internet, while older patients turn to more old-fashioned media such as DVDs. It can be analyzed that there is no variety of material developed for this audience, it was noticed that the material on social networks is not adequate, because it is old, made in a "homemade" way and with information that does not match the reality of the patients.

This project aims to provide access to this information through an audiovisual production. This production aims to collaborate in the treatment with exercises, in order to promote patient autonomy in a credible material and appropriate language. The exercises do not replace physiotherapy sessions, but complement them.

This document will present all the studies carried out for the formation of this project, from theoretical surveys, with bibliographic reviews, to the development of the project, also presenting the validation, which is the test with the end user.

GENERAL OBJECTIVE

To produce a series of five videos aimed at people with Parkinson's disease in order to give them autonomy in physiotherapy exercises.

SPECIFIC OBJECTIVES

- Identify and analyze the needs of the target audience and professionals in the field;
- Define which techniques and audiovisual language will be used in the project;
- Develop the pre-production, production and post-production process;
- Validate the project with the target audience and professionals in the field;
- Promote the project through social networks;

BACKGROUND

Parkinson's disease is caused by a progressive neurological alteration, which mainly affects the motor system due to a decrease in dopamine production. This disease mainly affects people over the age of 50 of both sexes (LANA, et all. 2007). A brief survey showed that there is not much material developed for this audience. And it was noticed that the material on social networks is not suitable, as it is done in a "home-made" way and contains a lot of information. The aim of this project is, through design, to develop audiovisual communication that is more accessible to these patients. Thus, this project should meet their needs so that they can perform some exercises at home. With the collaboration of specialists in the field, so that the information is reliable. This project will give patients independence again, with some simple exercises that can be done without supervision and will not expose the patient to any risk. The exercises do not replace physiotherapy sessions, but complement them.

Chapter 2

2. DATA COLLECTION

2.1 PARKINSON'S DISEASE

2.1.1 A brief history

The symptoms of Parkinson's were first identified scientifically in 1817 by the English physician Dr. James Parkinson in London, who observed and monitored the disease in a group of six patients. He noticed that the patients had similar symptoms, such as shaky arms and hands, slowness of movement, signs of muscle weakness and other symptoms. Parkinson concluded that this was a new disease and then published the article "Essay on Agitating Paralysis", which was the first record of the disease (REIS, 2004).

In 1860, in France, Jean-Martin Charoct, one of the leading neurologists of the time, became interested in Parkinson's article and decided to expand on the studies already carried out. Recognizing Parkinson's efforts and his originality, Charoct proposed naming the disease Parkinson's (REIS, 2004).

2.1.2 What is Parkinson's Disease?

According to Lana et al. (2007), Parkinson's is a progressive neurological disorder, which mainly affects the motor system due to a decrease in dopamine production. This disease mainly affects people over the age of 50 of both sexes.

> PD is characterized by motor disorders and postural dysfunctions. The main motor disorders are bradykinesia (slowness of movement), hypokinesia (reduced range of movement), akinesia (difficulty initiating movements), tremor and rigidity, as well as balance and gait deficits. As the disease progresses, patients may present cognitive disorders, memory deficits, problems related to visual-spatial dysfunction, difficulties in performing sequential or repetitive movements, freezing and slow psychological responses (LANA, et all. 2007).

All these symptoms can lead to speech and writing difficulties, and possibly depression. PD (Parkinson's Disease) is not fatal, nor is it contagious, and there are no indications that it is hereditary. In most cases, the Parkinson's sufferer's intellect remains intact. Another factor that can occur in some parkinsonians is the loss of facial expression, caused by the rigidity of the facial muscles, which makes it difficult to identify whether the patient is sad or happy (GROSSMANN, 1998).

> Taking into account statistics from other countries and considering that PD is a universal disease, it is likely that the number of parkinsonians in Brazil is around 300,000 patients. Worldwide, this figure could exceed 10 million people (REIS, 2004, p.179).

PD is a progressive disease and there is no cure. Furthermore, its progression is considered to be rapid due to the fact that patients become incapacitated in around 5 years, after which time the progression becomes a little slower and more prolonged (LIMA, 2008).

As soon as treatment is started quickly and intensively, the symptoms are alleviated and other complications are avoided, but over the years the drugs become less effective in their function (LIMA, 2008).

2.1.3 Treatment

During the First World War, there was an encephalitis pandemic that killed thousands of people. It was soon noticed that the survivors of encephalitis had symptoms that were similar to those of Parkinson's disease, but which did not have the same clinical characteristics. These symptoms and signs gave rise to Parkinsonism (REIS, 2004).

In 1921, with the collaboration of Tretiakoff, it was possible to define that there was a considerable loss in a region of the brain known as the "Black Substance".

This substance is located in the brain region called the midbrain and is responsible for producing dopamine (REIS, 2004).

In the 1950s, in Sweden, according to Reis, (2004) "[...] 'Dopamine' was a chemical substance fundamental to brain function, performing the function of a neurotransmitter [...]".

> In 1967/1968 Cotzias, in the United States, introduced L-Dopa, a precursor molecule to Dopamine, which revolutionized the treatment of PD. This substance is

<blockquote>biochemically able to reach the brain and transform into Dopamine there. The era of repositioning therapy began, an unprecedented victory in the treatment and improvement of PD (REIS, 2004, p.19).</blockquote>

L-Dopa was used as a treatment resource in 1968 and continues to be used today, as it is still the most widely used drug to help treat PD. However, neurologists and patients know that, like other drugs, L-Dopa begins to lose its effect after a while and also causes side effects (REIS, 2004).

Another means of treatment besides medication is surgery. It emerged in the 1950s, when treatment with L-Dopa (levodopa) had not yet been discovered. When surgeries had positive results, they brought great improvements to patients, but when they weren't effective, they brought certain complications to patients, such as blindness, coma and paralysis (GROSSMANN, 1998).

Nowadays, surgeries are more efficient, with better conditions and equipment, and consequently there has been a reduction in the risk of this procedure. However, the complications remain the same, and end up occurring in 1 out of every 10 surgeries (GROSSMANN, 1998).

2.1.   4Physiotherapy

According to Santos et al. (2010, p. 20) "Physiotherapy in PD was first established by Morris in 2000. He described specific strategies to improve activities such as walking, transfers, manual activities, preventing falls and maintaining physical capacity."

Physiotherapy is part of the treatment of PD and is indispensable for its sufferers, as it contributes to the effectiveness of the treatment (GROSSMANN, 1998). The aim of physiotherapy for people with Parkinson's is to reduce the motor problems caused by the first symptoms, helping the patient to remain independent and thus improving their quality of life (SANT, 2008). Physiotherapy sessions should include motor, walking, relaxation and breathing exercises. They should also re-educate patients and their families about the benefits of exercise (SANTOS et. all, 2010).

<blockquote>Physiotherapeutic intervention includes conventional and occupational therapy, therapy with visual, auditory and somatosensory stimuli. The stimuli facilitate movements, the start and continuation of walking, an increase in step size and a reduction in the frequency and intensity of freezing (SANTOS et. all, 2010, p.18).</blockquote>

The exercises given by physiotherapists aim to improve movements that were once commonplace, such as getting up, walking and reducing falls. Some authors consider the reduction of falls to be the most important, as 68.5% of patients with Parkinson's suffer falls, of which 33% have bone fractures. The aim of physiotherapy is to keep patients safe and prevent further complications (SANTOS et. all, 2010).

2.1.   5Speech therapy

According to Ferraz and Borges (2016): "Decreased vocal volume and dysarthria are frequent disorders that are not very responsive to drug treatment." This is why speech therapy plays an important role in the treatment of PD.

Because of their oral motor skills, PD patients may have difficulty formulating sentences and swallowing. Many patients feel ashamed because they can't express what they want. This is why speech therapy is important to help these patients who end up isolating themselves out of fear and embarrassment. Speech therapy aims to improve the patient's communication and eating habits. It stimulates the muscles that are responsible for these functions (REIS, 2004).

2.2 MEDICAL INFORMATION ON THE INTERNET

The growth in the use of the Internet has made it possible to access a wide range of information, including medical information. Patients first seek out this channel because it is confidential and inexpensive.

<blockquote>In recent years, the world has witnessed a major technological transformation that has substantially increased access to information, especially via the Internet. In the area of health, more and more information is available. Access to technical-scientific information, combined with an increase in the population's level of education, has led to the emergence of a patient who seeks information about their illness, symptoms, medication and the cost of hospitalization and treatment: the expert patient (GARBIN; NETO; GUILAM, 2008, p. 579).</blockquote>

While access to this information is innovative and positive, there is a complicating factor:

the lack of filtering of what reaches patients. A lot of false information or information of dubious origin circulates on the internet, as Garbin; Neto; Guilam (2008, p.580) warns.

2.3 AUDIOVISUAL LANGUAGE

Audiovisual language emerged as a way of reproducing reality and recording everyday events. According to Martin (2004), cinema has become a means of communication, propaganda and information. It has become increasingly common to use this language in media such as TV and the internet.

> Man has become an audiovisual being in the literal sense of the word. From the first moments of his life, he can communicate audiovisually with the outside world, by means of a videographic scan that shows him in his mother's womb and allows him to hear his heartbeat. Then their whole life is permeated by audiovisual technologies (ALVES; FONTOURA; ANTONIUTTI, 2008, p. 13).

However, in the case of technical and medical information, it is important to emphasize that quality is a considerable factor, since there are many options for productions on the web, but most of them are of poor quality.

> I don't mean to belittle the importance of the creators who, using precarious resources, keep the paths of creation open to new aesthetics and renew cultural production in the fields of cinema, books and poetry, but rather to consider the fact that the production of a film for television or cinema has become something so expensive and so complex that it is almost impossible to do it in an artisanal way, given that the technical quality is very expensive and sophisticated and also given that the public has once again become accustomed to demanding quality in order to accept the message (COMPARATO, 1995, p. 51).51).

This information can be confirmed by talking to professionals in the field, Patricia Barrichello Nazari[1][2][2] (2016) from the APPP and Daiane Cristine Martins Ronchi[2] (2016) from the physiotherapy course at Positivo University, who claim that the videos on the web today are out of touch with reality and teach things that are harmful to treatment.

2.4 YOUTUBE

Along with the spread of the internet, platforms for distributing and sharing audiovisual material have emerged. The main sites were Vimeo[3], Youtube[4], Myspace[5], among others. The site that has achieved the most visibility is Youtube. According to a 2015 survey carried out by Google Brasil in partnership with Provokers, and Brand Building on Mobile, with Research Reds, Youtube is the favorite of Brazilians. 1500 people were interviewed and it was found that 69% of Brazilians use the internet to watch videos, of which 95% use Youtube (THINK WITH GOOGLE, 2016).

> In April 2008, 11 billion videos were watched in the US by 71% of the country's internet users. They watched an average of 228 minutes. Most of this audience went to YouTube, with 4 billion videos watched (TELLES, 2009, p. 114).

It stood out because of its technological differential, offering no barriers to sharing content. Its interface was simple and made it possible to embed videos on other sites, such as blogs, as well as promoting interaction between users. Another favorable point of the tool is the user's autonomy to repeat the information as many times as they deem necessary, through the media playback controls.

> By the beginning of 2008, according to various web traffic measurement services, it was already consistently among the 10 most visited sites in the world... ComScore, an internet market research company, reported that the service accounted for 37% of all videos watched in the USA (BURGESS; GREEN, 2009, p.18).

Nazari (2016) says that patients often come home and spend hours on the internet, a fact that Telles (2009) describes with the following phrase: "YouTube is the TV of the digital generation".

Among patients, there is a need to use DVD media, which is the second way they watch videos. DVD (Digital Versatile Disc) is a digital data storage format. Users who don't have access to

---

[1] Physiotherapist at the Paraná Association of Parkinson's Disease Patients

[2] Physiotherapist and Master in Physiotherapy at Positivo University

[3] Vimeo, available at: https://vimeo.com/

[4] Youtube, available at: https://www.youtube.com/

[5] Myspace, available at: https://myspace.com/

the internet or digital media can access videos using this format on a DVD player.

2.5 AUDIOVISUAL PRODUCTION

Visual language is divided into three elements: The sender, the receiver and the medium in which it is transmitted. For Mccloud (2008), a production can be considered quality when the three elements are in harmony and the message is clear. For Kloter and Armstrong, these elements extend to nine: sender and receiver (main parts), tools, message and media, encoding, decoding, response and feedback (KLOTER AND ARMSTRONG apud ALVES et. All., 2008). Feedback is achieved through the audience, so the message has to be appropriate for those for whom it is intended.

> Within the approach of greater efficiency of communication and waiting for a previously planned return, it is necessary to consider that not all forms of communication involve grouping everyone together at a given time. It is therefore advisable to subdivide the masses, using different variables, alone or in combination... This is what we call segmenting consumer markets (ALVES; ANTONIUTTI; FONTOURA, 2008, P. 99).

The stages of a production according to Alves; Antoniutti; Fontoura (2008) consist of pre-production, production and post-production. The pre-production stage is the stage after the script has been approved, where the details of the production are agreed. Production is the stage where everything defined in pre-production will be put into practice, it's the moment of recording and filming.

2.5.1 Roadmap

According to Comparato (1995) "A good script is no guarantee of a good film, but without a good script there is certainly no such thing as a good film [...] A good script can be summed up in a sentence that encompasses conflict." Comparato also explains the phases of a script. The first step is to define a theme, from which an idea for the production will emerge. The following steps may or may not fit into a production like the one in the project, which would be the choice of a character, the dramatic time and the dramatic unit. The character is the one who will convey the message, in this case. They are the ones who will give the viewer confidence or even identification. The dramatic unit is the part where we define the sequence of scenes, i.e. the storyboard. Dramatic time is where the project's execution time is defined. The author gives as an example the viewer's attention span on TV, which would be 3 minutes.

2.5.2 Vignette

In audiovisual productions broadcast on the YouTube network, the use of opening and closing vignettes has become increasingly common. This is the moment when the visual identity of the project is presented or reinforced, along with the credits to the sponsors.

> It is shown before the start of a program or event and/or immediately after it ends, to identify the sponsor. It should be 7 seconds long, like a movement, and have a maximum of twelve words, and it is essential to highlight the advertiser right at the start. (Alves; Antoniutti; Fontoura, 2008, p.159)

In addition to being present on the web, the vignette has long been part of media such as TV, cinema and DVD, and is widely used in advertising with commercials.

FIGURE 01 - SCREENSHOT OF NAMU FIT CHANNEL VIGNETTE

SOURCE: https://www.youtube.com/watch?v=3sFz4dtcxJs

2.5.3 Production Elements

McCloud (2008) says that the basic aim when telling a story is for the listener to understand what is being said and to be hooked enough to continue. A clear sequence is one that contains only what is necessary to convey the message. Other aspects can also contribute to this objective, for example, the choice of frame transitions. McCloud (2008) gives six types as examples: moment to moment (an action represented in a series of moments), action to action (a character in a series of acts), subject to subject (a series of characters in a single scene), scene to scene (transitions between time and space), aspect to aspect (transitions of aspect that give a sense of place, idea or state of mind), non sequitur (series of words and images with no connection). In the case of this production, the moment-to-moment transition is considered the most appropriate, as it focuses on what is being shown and accompanies the development in a more focused way.

Another aspect of the production will be the framing. The closer the shot is to the action, the greater the attention to detail, but if the aim is also to provide a sense of integration between the production environment and the viewer, there should be a step back. McCloud (2008) points out that keeping the surroundings free of major changes helps to keep the viewer's attention fixed on the area where the change is important. Normally, importance is given to objects and characters positioned in the center of the frame. Comparato distinguishes between foreground (close up or detail shot), medium shot (cuts the figure at the waist), American shot (cuts the figure at the knee) and general shot (emphasis on the environment).

FIGURE 02 - EXAMPLE OF FOREGROUND IN SCREEN CAPTURE OF EXCERPT FROM "LOVE AND OTHER DRUGS", A MOVIE ABOUT PARKINSON'S DISEASE

SOURCE: https://www.youtube.com/watch?v=DVJ2mcREl2M Accessed on June 12, 2016

FIGURES 03 AND 04 - EXAMPLES OF MEDIUM SHOT AND AMERICAN SHOT, RESPECTIVELY, TAKEN FROM A SCREEN CAPTURE OF THE MOVIE "LOVE AND OTHER DRUGS".

SOURCE: https://www.youtube.com/watch?v=DVJ2mcREl2M
Accessed on June 12, 2016

FIGURE 05 - EXAMPLE OF A GENERAL SHOT TAKEN FROM A SCREENSHOT OF THE MOVIE "LOVE AND OTHER DRUGS".

SOURCE: https://www.youtube.com/watch?v=DVJ2mcREl2M
Accessed on June 12, 2016

2.5.4 Post-Production

Post-production is when the material collected is finalized or edited.

> With the raw material in hand, the director, together with the editor/finisher, will begin to assemble the film/video, which consists of choosing the scenes, organizing the sequence, working on the effects, until the material is the way he imagined it when he read the script (ALVES; ANTONIUTTI; FONTOURA, 2008, P.185).

This is the final part of a production, editing is important because it defines what is important in the final product and what should be left out, as well as audio and sound effects that could not be captured in the production stage.

2.6 DATA COLLECTION

Data was collected in two different situations. One was at the Paraná Association of Parkinson's Disease Patients, through a structured interview, and the other was carried out through an online questionnaire in groups aimed at Parkinson's disease patients.

Based on the data collected, it was possible to determine the age range of the target group, which is between 40 and 86 years old. Many of these patients discovered the disease a few months ago, while others have been living with the disease for 10 years. Despite the difficulty in doing some exercises, the vast majority know that physiotherapy can give them better living conditions.

The data collected revealed a difference between these audiences in terms of access to information. As can be seen in the following information.

When asked if they usually access the internet, 91.7% of APPP patients do not have access to the internet and many said they were not interested in the subject, while 85.7% of those who answered the questionnaire online do, as can be seen in the following graphs:

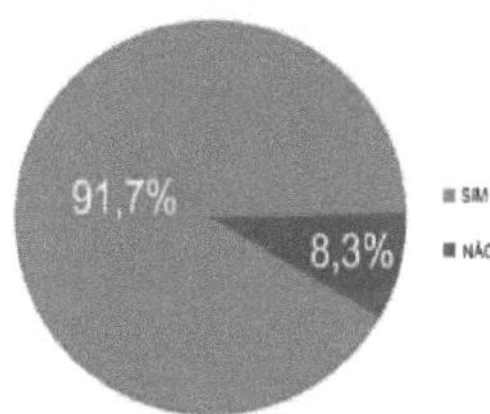

13

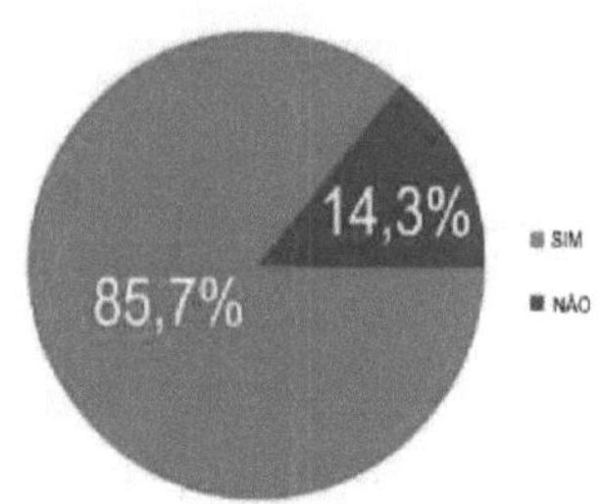

SOURCE: PREPARED BY THE AUTHORS BASED ON THE QUESTIONNAIRE

They were also asked if they had a DVD player, 66.7% of the APPP patients said they didn't have one, while 75% of the online public said they had a DVD player, as shown in the graphs below:

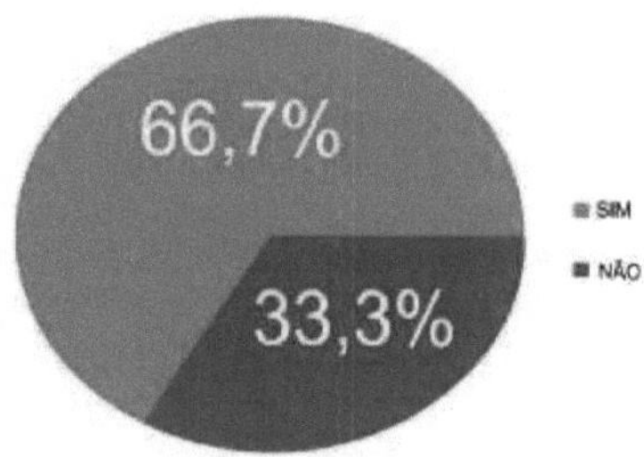

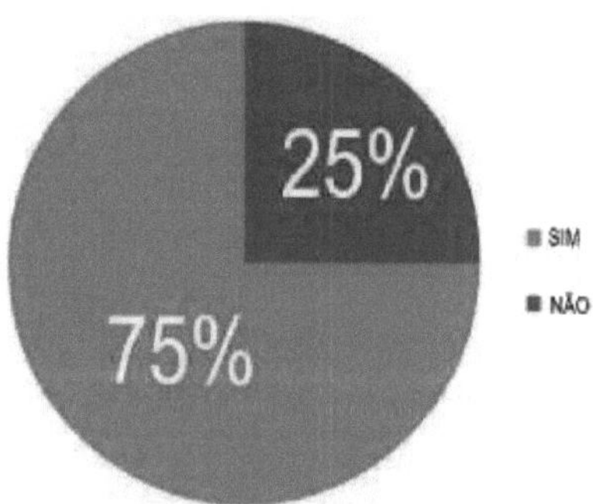

These responses led to the need to find out in which medium they would like to have access to the videos that will be developed in this project. At APPP 58.3% of patients would like to watch the videos on Smart TV and 41.7% on a computer, with the help of a family member. As for the online audience, 67.9% would like to watch on a laptop and 46.4% would like to watch on a desktop computer, as shown in the graphs:

GRAPH 5 - RESPONSE FROM APPP PATIENTS

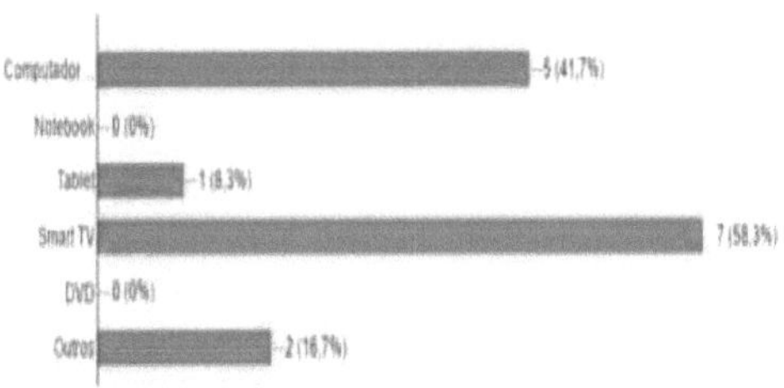

GRAPH 6 - RESPONSE FROM PARKINSONIANS ONLINE

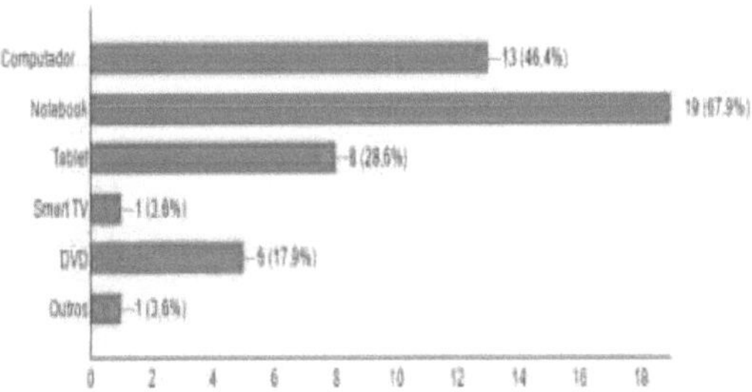

Although no questions were asked about the income of those interviewed, it was noted that the patients at the APPP belong to classes C and D, while those who answered the questionnaire online belong to classes A and B. Therefore, it is possible to see that this project will have to cater for these two different audiences, in such a way that neither of them lacks information. Because the media that will access this content are also different.

2.6.1 Personas

After research to define the target audience, we noticed the need for a tool to help visualize and understand this audience during the development of the project. To this end, the persona creation tool was used.

A persona is a fictional character who presents behavioral patterns based on the characteristics of a target audience. It is a generalized representation of a group. In this project, we saw the need to cater for two target audiences: the patients of the APPP and Parkinson's sufferers who were contacted via social networks. Based on this, two personas were defined to represent these audiences.

SOURCE: PREPARED BY THE AUTHORS

The first persona represents the APPP's audience, who are people of advanced age, with little or no access to the internet and at an advanced stage of the disease.

FIGURE 7 - ONLINE PERSONA

SOURCE: PREPARED BY THE AUTHORS

The second persona represents the social media audience, people who are on average 40 years old, discovered the disease at an early stage and use the internet as a means of obtaining information about the disease.

2.7 ANALYSIS OF DIRECT SIMILARITIES

In order to determine how the project's existing counterparts are produced, existing YouTube videos, direct and indirect counterparts will be analyzed. Graphic and production characteristics will be observed, such as: the existence of a vignette, its duration, its aesthetics, the type of font used, the information it contains and the colors used; the visual identity of the video, the colors and elements used; whether it is possible to see that there was a script for the video; the scenery presented, whether there is any graphic interference during viewing, the main character (professional, patient or both), the fonts and colors used in the video, the lighting, the audio quality, whether there is a soundtrack, how the scenes are transposed, which shot is used, whether there are any camera effects, whether the language is appropriate (is it suitable for the public or does it use technical and/or scientific terms), the theme used, the language, the number of views on the channel, the length of time it lasted and the year it was published.

With this information, we want to determine the graphic quality of the videos that already exist, the size of the demand for them, and whether there is any credible information published recently.

2.7.1 Canal Weder Alves da Silva

YouTube channel with recordings of exercises carried out in a gym, with a Parkinson's sufferer following instructions from her Personal Trainer.

FIGURES 08 and 09 - SCREEN CAPTURE OF THE WEDER ALVES DA SILVA CHANNEL

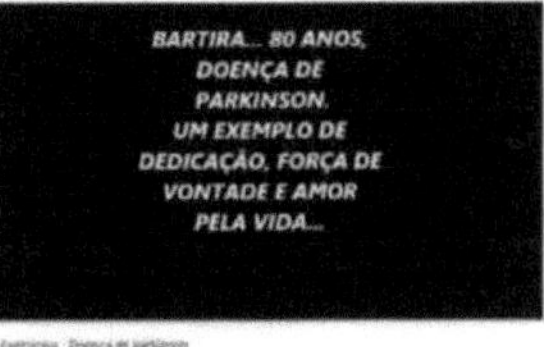

SOURCE: https://www.youtube.com/watch?v=vXu9uMye48U Accessed on June 05, 2016

Pre-production

- **Vignette:** None. The video begins with a statement in white font on a black background. Large capital letters.
- **Visual Identity:** None.
- **Script: There** is no indication that a script was planned, as the exercises are presented in no specific order, and it appears to be a record of a training session rather than a tutorial.

Production/Post-production

- **Scenario:** Training ground, gym with equipment on display.
- **Graphic interference:** None.
- **Characters:** Personal trainer and Parkinson's student. There is no one in the background.
- **Fonts:** None throughout the video, only in the opening and closing.
- **Colors:** Only the black and white of the opening and closing. There is no color editing in the video.
- **Lighting:** Room lighting only, change from external to internal only.
- **Audio:** No dialog, just a soundtrack.
- **Soundtrack:** Pop song, Katy Perry - Fireworks. The song plays for part of the video and then cuts out without ending and repeats itself, without compromise.
- **Transition of scenes:** Additive dissolution effect on scenes.
- **Filming shots:** General shot.
- **Camera effects:** There aren't any, you can see that someone is holding the camera and following the exercises.
- **Use of technical terms:** No dialogue.

- **Language:** The channel is Brazilian, the opening and closing texts are in Portuguese.
- **Topic in focus:** Exercises for Parkinson's sufferers, led by a personal trainer.
- **Number of views on the video:** 19.103
- **Running time:** 00:08:18
- **Year published:** 2013

2.7.2 Alice Pinho Channel

YouTube channel showing examples of various exercises for Parkinson's sufferers, developed by the University of Lisbon, Portugal.

FIGURE S 10, 11, 12 AND 13 - ALICE CHANNEL SCREENSHOT

SOURCE:https://www.youtube.com/watch?v=Y1kSn1HCKk8
Accessed on June 05, 2016

Pre-production

- **Vignette:** None. The video begins with an image identifying members of the University of Lisbon in Portugal.
- **Visual identity:** None. There is a brand next to the opening and closing image with the words molecular medicine, but it doesn't appear to be from the video itself, as it isn't highlighted.
- **Script:** There doesn't seem to have been a script for capturing the videos, but there is an order to the editing sequence.

Production/Post-production

- **Scenario:** Clinic or gym, with equipment.
- **Graphic interference:** Exercise titles appear as the video changes.
- **Characters:** Parkinson's patients.
- **Fonts:** You can see the concern for contrast in the change of colors between white, yellow and black. But reading is impaired in some cases where the font occupies the space of a device present in the footage, with the same color as the font. There is also a change in the position of the information, with the font appearing at the bottom of the video most of the time, but sometimes at the top.
- **Colors:** Only the font colors (black, yellow and white) and the image at the beginning is mostly red and blue.
- **Lighting:** Room lighting only.

- **Audio:** No dialog, just a soundtrack.
- **Soundtrack:** Better Not Stop, uncut.
- **Transition of scenes:** "Dry" cut.
- **Filming shots:** General shot.
- **Camera effects:** There aren't any, you can see that someone is holding the camera and following the exercises.
- **Use of technical terms:** There are some in the exercise legends.
- **Language:** The channel is in Portuguese, but the texts are in English.
- **Topic:** Examples of physical, sensory and cognitive stimulation exercises for Parkinson's Disease.
- **Number of views on the video:** 37.986
- **Running time:** 00:04:25
- **Year published:** 2012

2.7.3 Canal   Revista da Cidade

Report taken from the city magazine YouTube channel. It talks about the importance of physiotherapy in treatment and gives examples of exercises to do at home.

FIGURES 14 AND 15 - SCREENSHOT OF THE CITY MAGAZINE CHANNEL

SOURCE: https://www.youtube.com/watch?v=Smg1M-sR8ek
Accessed on June 05, 2016

Pre-production
- **Vignette:** There is no **vignette** because the video starts at a certain point in the original program.
- **Visual Identity:** It has that of the program, the brand is visible throughout the video in the bottom left corner.
- **Script:** Yes, you can see from the questions asked by the presenter and the physiotherapist that she has a simple order of exercises.

Production/Post-production
- **Set:** TV studio with makeshift bed and chair simulating places you'd find at home.
- **Graphic interference:** Only the title of the story, which appears and disappears as the video progresses.
- **Characters:** Program presenter, physiotherapist and patient.
- **Fonts:** Readable font in the title, no problems with contrast.
- **Colors:** Cool tones. Shades of blue and gray.
- **Lighting:** Studio lighting.
- **Audio:** Studio-quality audio.
- **Soundtrack:** Soft ambient music in the background.
- **Transition of scenes:** None.
- **Filming shots:** General shot, medium shot, American shot and foreground.
- **Camera effects:** You can see at least two cameras in different positions different shots, which makes the video more dynamic.
- **Use of technical terms**   : There is concern about understanding the spectator, uses simple terms.
- **Language:** The channel is Brazilian   , all the dialog is in Portuguese.
- **Topic in focus:**   Examples of functional and physiotherapeutic exercises homemade.

- **Number of views on the video:** 54.891
- **Running time:** 00:08:55
- **Year published:** 2012

Neurology Channel 21

Video from the Neurology 21 channel on YouTube. It has several examples of exercises led by Pamela Quinn, a professional dancer and choreographer with the disease.

FIGURES 16, 17, 18 AND 19 - NEUROLOGY CHANNEL SCREENSHOTS

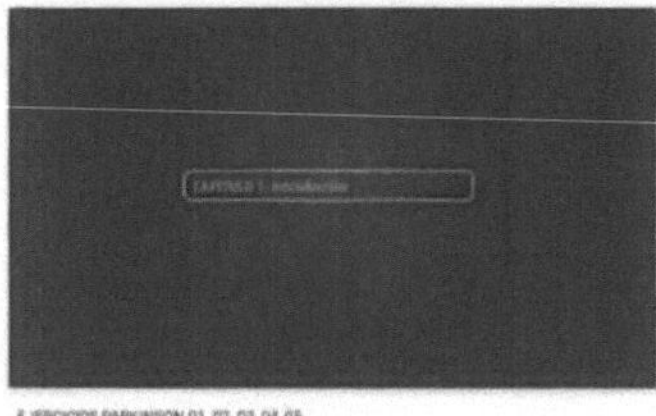

SOURCE: https://www.youtube.com/watch?v=T-zmzuSGRWc
Accessed on June 05, 2016

Pre-production

- **Vignette:** Has.
- **Duration:** 00:00:06
- **Style**: Color frame with audio and text information appearing at the beginning of the video and in all chapter transitions.
- **Font**: It has contrast, but it's small compared to the size of the picture.
- **Information presented**: Presentation of the chapter and what it is about.
- **Colors**: Pink background, salmon font, ellipse with white outline.
- **Visual Identity:** None.
- **Script:** Yes, you can tell by the way the video is conducted. The order of the exercises is programmed.

Production/Post-production

- **Scenario: It** looks like the living room of a house, but it's empty. It only has chairs, curtains and plants.
- **Graphic interference:** None.
- **Characters:** Pamela Quinn presenting the exercises (dancer, choreographer and patient), and two other Parkinson's patients accompanying the exercises.
- **Fonts:** Only in the vignette.
- **Colors:** Colors of the vignette, and in the video, there doesn't seem to be any color editing, but the environment is brown, leaving the whole video in shades close to color.
- **Lighting:** Ambient lighting and there seems to be studio lighting too.
- **Audio:** Good quality audio, but dual language.
- **Soundtrack:** Ambient music one tone below the audio with instructions, at background.
- **Transitioning scenes:** The transition is made by the vignette.
- **Filming shots:** General shot, medium shot, American shot and foreground.

- **Camera effects:** There are at least four cameras capturing images. It seems that one is frontal and static. The others follow the movements, one on the left, another on the right, and the last one is positioned above or below the characters.
- **Use of technical terms:** As far as I can tell, the vocabulary is simple and easy for patients to understand.
- **Language:** The original audio is in English, but there is a dubbing in Spanish.
- **Topic in focus:** Examples of exercises for Parkinson's sufferers.
- **Number of views on the video:** 163.166
- **Running time:** 00:46:39
- **Year published:** 2013

2.7.5 Daiane Seibert Channel

Video simulation of rehabilitation training for people with pathologies, in this case Parkinson's, made by physical education and nursing students from the Lutheran University of Brazil - Gravataí.

FIGURES 20, 21, 22, 23 AND 24 - SCREENSHOTS FROM DAIANE SEIBERT'S CHANNEL

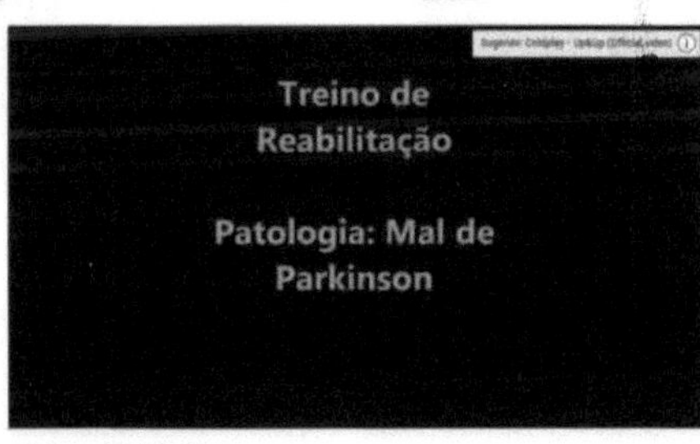

Treino para Mal de Parkinson

SOURCE:https://www.youtube.com/watch?v=6DqWLe7OAQE
Accessed on June 05, 2016

Pre-production
- **Vignette:** None. At the beginning and end of the video there is a statement with a black background and white font with information about the video.
- **Visual identity:** None.
- **Script:** There was planning in the sequence of exercises.

Production/Post-production
- **Scenario:** Outside the college, next to a sports court.

- **Graphic interferences:** Texts indicating which exercise is being done.
- **Characters:** Four students, one playing a Parkinson's patient and the other three as professionals.
- **Fonts:** They appear during the video above the images, in yellow and red, as well as in the statement. During the video, they are large, sometimes taking up the entire frame.
- **Colors:** Black, white, red and yellow for fonts and opening backgrounds. The video has not been edited.
- **Lighting:** Ambient lighting.
- **Audio:** At the beginning there is a dialog section with a lot of ambient noise. Another problem is that the music chosen for the soundtrack is louder than the audio.
- **Soundtrack:** POP songs: Coldplay and Black Eyed Peas.
- **Scene transition:** "Dry" cut.
- **Filming shots:** General shot and medium shot.
- **Camera effects:** One person follows the movements, filming is irregular.
- **Use of technical terms:** Terms from the students' area are used, such as performance scales.
- **Language:** All in Portuguese.
- **Topic in focus:** Exercises from a rehabilitation program for Parkinson's patients.
- **Number of views on the video:** 2,505
- **Running time:** 00:11:06
- **Year published:** 2014

2.7.6 Caio Vicentini Channel

A video found on YouTube shows physical education exercises adapted for elderly people with Parkinson's disease.

FIGURES 25, 26, 27, 28 AND 29 - SCREENSHOT CAIO VICENTINI CHANNEL

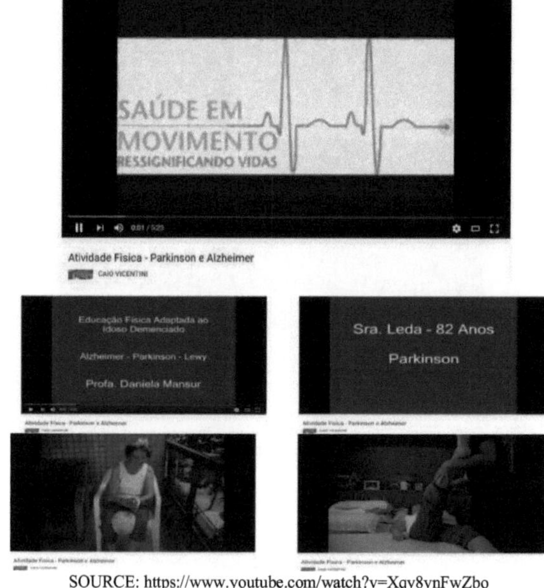

SOURCE: https://www.youtube.com/watch?v=Xqy8vnFwZbo
Accessed on June 05, 2016

Pre-production
- **Vignette:** Has.
- **Duration:** 00:00:04
- **Style:** Brand image with retro video effect (monochrome film with noises).
- **Font:** Only the one in the image, which is legible.
- **Information provided:** Brand.
- **Colors:** Monochrome.
- **Visual Identity:** The brand that only appears in the vignette.
- **Script:** There was an order for the exercises.

Production/Post-production
- **Scenario:** Patient's home.
- **Graphic interference:** During the video none. There are texts between each illness identifying the patient and their case.
- **Characters:** Patient and professional.
- **Fonts:** Only when changing patients, white with blue background in good size.
- **Colors:** Only in transigoes with titles, blue.
- **Lighting:** Ambient lighting.
- **Audio:** You can hear the professional's instructions in the background, but you can't make out what she's saying; the soundtrack stands out.
- **Soundtrack:** Instrumental, standing out from the audio. No compromise.
- **Transition of scenes:** "Dry" cut.
- **Filming shots:** General shot.
- **Camera effects:** In one part the professional films herself, and in another the camera is static, with no effects.
- **Use of technical terms:** No dialog.
- **Language:** The texts are in Portuguese.
- **Topic in focus:** Physical education exercises for Parkinson's, Lewi's and Alzheimer's patients.
- **Number of views on the video:** 18.860
- **Running time:** 00:05:23
- **Year published:** 2012

2.7.7 The Barbara C. Dias

Video with examples of functional physiotherapy exercises for the elderly to do at home.

FIGURES 30, 31, 32 AND 33 - SCREENSHOTS FROM THE BARBARA C. DIAS CHANNEL

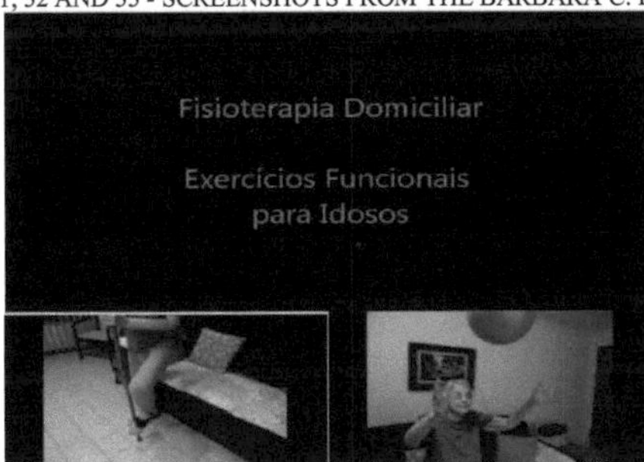

SOURCE: https://www.youtube.com/watch?v=Bgl0Aan1aoo
Accessed on June 05, 2016

Pre-production
- **Vignette:** None. At the beginning there is a title in white with a black background describing the subject of the video, and at the end the professional's details.
- **Visual identity:** None.
- **Script: It** doesn't show that there was, the exercises are records that are interspersed between one and the other.

Production/Post-production
- **Scenario:** Patients' home.
- **Graphic interference:** None.
- **Characters:** Patients.
- **Fonts:** Only at the opening and closing of the video, they are legible.
- **Colors:** No editing.
- **Lighting:** Ambient lighting.
- **Audio:** No dialog.
- **Soundtrack:** Electronic music "R.I.O - One Heart". In the comments there was a lot of praise for the soundtrack, but one elderly woman complained that she couldn't do the exercises with audio.
- **Scene transition:** "Dry" cut.
- **Filming shots:** General shot, American shot, Medium shot and Foreground.
- **Camera effects:** All sections are filmed by the professional accompanying the exercise.
- **Use of technical terms:** No dialog.
- **Language:** The texts are in Portuguese.
- **Topic in focus:** Functional physiotherapy exercises for the elderly to practice at home.
- **Number of views on the video:** 81.610
- **Running time:** 00:02:54
- **Year published:** 2013

2.7.8 Channel 1Parkinson

Video found on YouTube with exercises for Parkinson's, looking very homemade.

FIGURES 34, 35 AND 36 - SCREENSHOT OF CHANNEL 1PARKINSON

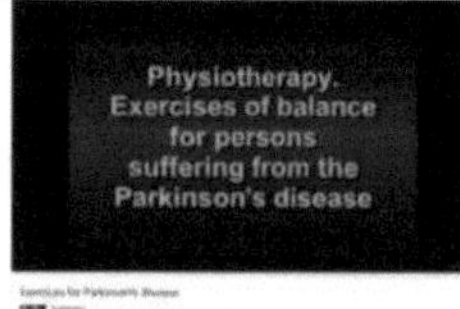

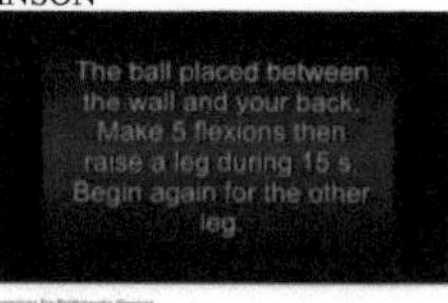

SOURCE: https://www.youtube.com/watch?v=n9ZZmK6YMXI
Accessed on June 11, 2016

Pre-production
- **Vignette:** None. At the beginning there is information in yellow text with a background in a gradient of blue and black.
- **Visual identity:** None.
- **Script:** Planning in the order of exercises only.

Production/Post-production
- **Scenario:** Clinic or gym, as you can see exercise equipment in the background.
- **Graphic interference:** There are not only texts announcing the next exercise.
- **Characters:** Patient.
- **Fonts:** On information boards, too large. Yellow with blue and black gradient background.
- **Colors:** No editing. Video looks old, very homemade.
- **Lighting:** Ambient **lighting.**
- **Audio:** No dialog.
- **Soundtrack:** Instrumental music.
- **Transition of scenes:** "Dry" cut. Generally, the cut is made with the entry of the frame with text.
- **Filming shots:** General shot.
- **Camera effects:** All sections are filmed by someone accompanying the exercises.
- **Use of technical terms:** No dialog.
- **Language:** All texts are in English.
- **Topic in focus:** Balance exercises for people with Parkinson's.
- **Number of views on the video:** 69.308
- **Running time:** 00:05:52
- **Year published:** 2009

2.7.9 Jean Paul de Cremer Channel

Video with information from professionals about the disease, showing not only examples of exercises to do at home, but also examples of what is done in clinics, including treatments with games such as "JUST DANCE".

FIGURES 38, 39 AND 40 - SCREENSHOT JEAN PAUL DE CREMER CHANNEL

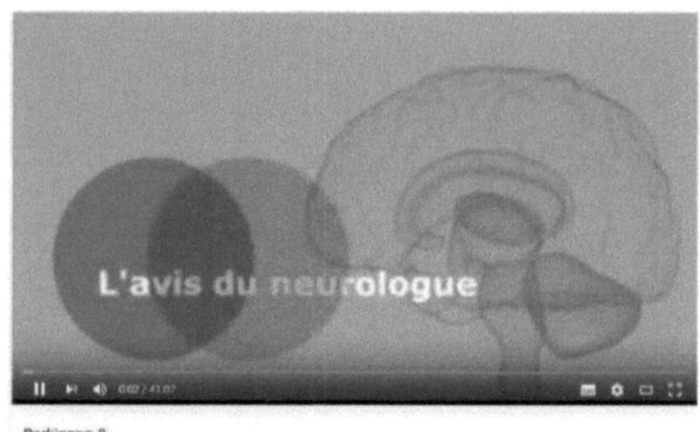

Parkinson 3

SOURCE: https://www.youtube.com/watch?v=cKbTQ-xugts
Accessed on June 11, 2016

Pre-production

- **Vignette: This** has a more elaborate opening than the previous examples, with a title and animated shapes, but the audio of the video itself comes with the opening.
- **Visual identity:** None.
- **Script:** There is. The whole video, both in terms of the subjects and the exercises, has a remarkable structure.

Production/Post-production

- **Scenario:** In the materials, the exercises are done in a clinic, in the examples it's a room, you can't tell if it's a house or a clinic. At the end there are some tips for home, so the scenario is a house.
- **Graphic interference:** Only when changing exercises is there a countdown to the start, and the caption of what it is about.
- **Characters:** Professionals and patients.
- **Fonts:** White and not interfering with the image, but small.
- **Colors:** The exercise transitions are done with white elements, but in the video there doesn't seem to be any color editing. You can see that the elements in the image make the video more colorful.
- **Lighting:** Ambient **lighting.**
- **Audio:** In the dialog parts, the audio is good, without noise, some parts have ambient music in the background, others don't.
- **Soundtrack:** Instrumental, upbeat music.
- **Scene transitions:** fade-in and fade-out cuts.
- **Filming shots:** General shot, Medium shot, American shot and First shot plan.
- **Camera effects:** There is at least one static **camera** and one moving **camera**, alternating shots.
- **Use of technical terms:** There seem to be some in the article, but as there are no dialogues in the examples, the text is simple.
- **Language:** French.
- **Focus:** Information about the disease and examples of exercises to do at home.
- **Number of views on the video:** 46.686

- **Running time:** 00:41:07
- **Year published:** 2013

2.7.10 John Argue Channel

This video is a record of a treatment applied to people suffering from Parkinson's with dance, yoga and tai chi activities.

FIGURES 41, 42 AND 43 - SCREENSHOTS JOHN ARGUE CHANNEL

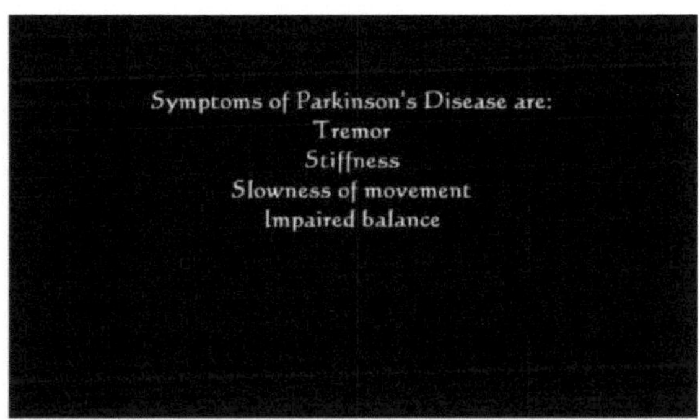

Parkinson's Dance: People with Parkinson's Can't Do That...Can They?!
John Argue

SOURCE: https://www.youtube.com/watch?v=pL_LZgAEsnM
Accessed on June 11, 2016

Pre-production
- **Vignette: There is** none, the opening is a text presenting the symptoms of the disease and the average time that the patients in the video discovered they were carriers.
- **Visual identity:** None.
- **Script: There doesn't** seem to be one, the video follows the order in which the exercises were performed.

Production/Post-production
- **Scenario:** Room, probably in a clinic.
- **Graphic interference:** None.
- **Characters:** Patients.
- **Fonts:** White on a black background, but ornate.
- **Colors:** No editing.
- **Lighting:** Ambient **lighting.**
- **Audio:** The dialog is good, with no interference or noise. The soundtrack has been reduced in volume for better understanding of the audio.
- **Soundtrack:** Instrumental, upbeat music.
- **Transition of scenes: The** effect of dissolving scenes.
- **Filming shots:** General shot and foreground.
- **Camera effects:** There is no fixed camera.
- **Use of technical terms:** Cannot be evaluated.

- **Language:** English.
- **Topic in focus:** Alternative exercises in the treatment of Parkinson's.
- **Number of views on the video:** 126.339
- **Running time:** 00:05:38
- **Year published:** 2011

2.7.   11Captainpat channel

The video shows the exercise routine of Neil Sligar, who was diagnosed with Parkinson's in 1998. The video was made in 2011.

FIGURES 44, 45 AND 46 - CAPTAINPAT CHANNEL SCREENSHOT

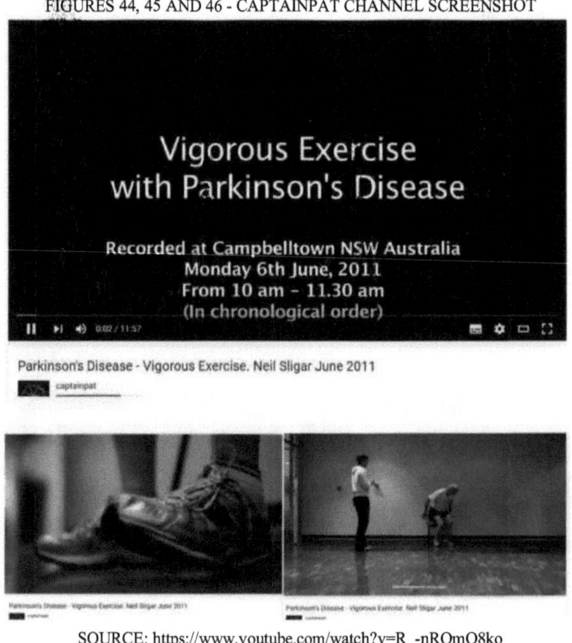

SOURCE: https://www.youtube.com/watch?v=R_-nRQmO8ko
Accessed on June 11, 2016

Pre-production
- **Vignette:** None, the opening is a text introducing the patient.
- **Visual identity:** None.
- **Script:** The opening says that the video follows chronological order.

Production/Post-production
- **Scenario:** Academy.
- **Graphic interference:** None.
- **Characters:** Patient and Professional.
- **Fonts:** White on a black background.
- **Colors:** The whole video tends towards shades of red, whether due to the lighting or the editing.
- **Lighting:** Ambient lighting.
- **Audio:** The part where there is dialog is good, without interference or noise, when there is music in the background, it doesn't interfere with understanding the audio.
- **Soundtrack:** Instrumental, soft.
- **Scene transition:** Dry cut.
- **Filming shots:** General shot, medium shot, American shot and foreground.
- **Camera effects:** Use overlapping frames, with a close-up superimposed on a general shot.
- **Use of technical terms:** Apparently not, as the professional talks to the student all the time.

- **Language:** English.
- **Topic in focus:** Monitoring the physical training routine of a Parkinson's patient.
- **Number of views on the video:** 46.814
- **Running time:** 00:11:57
- **Year published:** 2011

2.8 ANALYSIS OF INDIRECT SIMILARITIES

2.8.1 Canal Home Fit Home

This channel shows videos of exercises that you can do on your own at home. They are taught by a student who is training as a Health Coach, Personal Trainer and Pilates Instructor.

FIGURE 47, 48 AND 49 - SCREENSHOT OF THE HOME FIT HOME CHANNEL

SOURCE: https://www.youtube.com/watch?v=wbg43E2gVx4 Accessed June 14, 2016

Pre-production
- **Vignette:** No, at the beginning of the video there is a brief introduction to what the video is about, which is explained by the presenter.
- **Visual Identity:** Has.
- **Script:** You can see that there is a script, as the presenter sets an exact time for each exercise.

Production/Post-production
- **Scenario**: It's a house, but it's a very clean environment, with no objects that could take the focus off the exercise.
- **Graphic interference:** None.
- **Sources:** There is no text in the video.
- **Colors: The** environment is light, the colors that stand out the most are the objects used for the exercises and the presenter's clothes.
- **Lighting:** Ambient.
- **Audio:** You can hear well without any noise.
- **Soundtrack:** None.
- **Scene transitions:** General shots and American shots.
- **Camera effects:** Alternating angles.
- **Use of technical terms:** Use only the names of the exercises.
- **Language:** Portuguese.
- **Topic in focus:** Home training.

- **Number of views on the video:** 323,383 views.
- **Running time:** 00:17:31.
- **Year published:** 2015.

2.8.2 Fernando Gonçalves Channel

This channel focuses on aerobic exercise to burn calories even faster.

FIGURE 50, 51 AND 52 - SCREENSHOT OF THE FERNANDO GONÇALVES CHANNEL

SOURCE: https://www.youtube.com/watch?v=Hjwxj2M0XFU
Accessed on: June 14, 2016

Pre-production
- **Vignette:** Has.
- **Visual Identity:** Has.
- **Script:** There is a script, all the exercises are given step by step.

Production/Post-production
- **Setting:** It's a studio, with light colors.
- **Graphic interference:** None.
- **Sources:** None.
- **Colors:** The colors are neutral, such as gray, black and white.
- **Lighting:** Artificial.
- **Audio:** A little difficult to understand when there is dialog, as the music from the soundtrack interferes a lot.
- **Soundtrack:** There is, they're very upbeat songs.
- **Transitioning scenes:** This is done by changing shots, such as the American shot, general shot and focus.
- **Camera effects:** Alternating angles.
- **Use of technical terms:** None.
- **Language:** English.
- **Topic in focus:** Short aerobic exercises.
- **Number of views on the video:** 2,677,920 views.
- **Running time:** 00:29:56.
- Year published: 2014

2.8.3 Be Fit Channel

This channel has some video lessons that teach Pilates.

FIGURE 53, 54 AND 55 - SCREENSHOT OF THE BE FIT CHANNEL

Denise Austin: Abs & Core Pilates Workout

SOURCE: https://www.youtube.com/watch?v=UdtNcQteaOs
Accessed on: June 14, 2016

Pre-production
- **Vignette:** None.
- **Visual Identity:** Has.
- **Script:** Has.

Production/Post-production
- **Scenario:** Pilates studio.
- **Graphic interference:** None.
- **Sources:** None.
- **Colors:** Warm colors, both in the scenery and the clothes.
- **Lighting:** Ambient.
- **Audio:** Although the music continues in the dialog, it doesn't interfere with understanding it.
- **Soundtrack:** Yes, it's calm music.
- **Transitioning scenes:** Alternating shots.
- **Camera effects:** Alternating angles.
- **Use of technical terms:** Only the names of the exercises.
- **Language:** English.
- **Topic in focus:** Pilates.
- **Number of views on the video:** 191,507 views.
- **Running time:** 00:15:18.
- **Year published:** 2012.

2.8.   4Home Exercise Channel

The focus of the videos on this channel is a series of exercises that can be done at home.

FIGURES 56, 57, 58 AND 59 - SCREENSHOTS FROM THE EXERCICIOEMCASA CHANNEL

SOURCE: https://www.youtube.com/watch?v=UwW586G8GnU
Accessed on: June 14, 2016

Pre-production
- **Vignette: There isn'**t one, just a warning about injuries and some information about the class.
- **Visual Identity:** Has.
- **Script:** Has.

Production/Post-production
- **Scenery: There isn'**t any, it's just a white background.
- **Graphic interference:** None.
- **Fonts:** Legible, no problems at all.
- **Colors:** There's only green at the beginning of the video.
- **Lighting:** Artificial.
- **Audio: There's** a lot of interference between the music and the dialog.
- **Soundtrack:** There is, they're very upbeat songs.
- **Scene transitions:** None.
- **Camera effects:** Static camera.
- **Use of technical terms:** Only the names of the exercises.
- **Language:** Portuguese.

- **Topic in focus:** Exercises to do at home.
- **Number of views on the video:** 1,841,126 views.
- **Running time:** 00:17:35.
- **Year published:** 2014.

2.8.5 Namu Fit Channel

On this channel, the objectives of the videos are also exercises to be carried out at home.

FIGURE 60, 61,62, 63 AND 64 - SCREENSHOT OF THE NAMU FIT CHANNEL

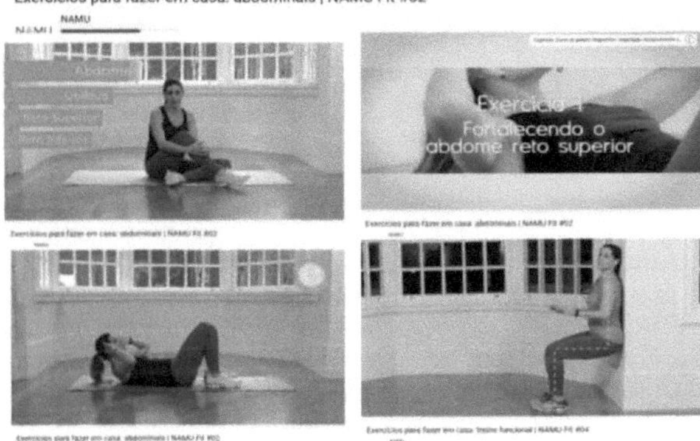

SOURCE: https://www.youtube.com/watch?v=kmxNHm6uv94
Accessed on: June 14, 2016

Pre-production
- **Vignette:** Has.
- **Visual Identity:** Has.
- **Script:** Has.

Production/Post-production
- **Scenario:** A house, with no distracting objects.
- **Graphic interference:** It does, and they are well resolved.
- **Fonts:** Legible, easy to understand.
- **Colors:** light, cool and neutral.
- **Lighting:** Ambient.
- **Audio:** Quality and easy to understand.
- **Soundtrack:** Yes, it's very subtle.
- **Transitioning scenes:** Alternating shots.
- **Camera effects:** Alternating angles.
- **Use of technical terms:** None.
- **Language:** Portuguese.

- **Topic in focus**: Exercises to do at home.
- **Number of views on the video:** 256,449 views.
- **Running time:** 00:14:23.
- **Year published:** 2015.

TABLE 1 - SUMMARY OF THE ANALYSIS OF DIRECT AND INDIRECT SIMILARITIES

Canal/ano	PRÉ-PRODUÇÃO			PRODUÇÃO					PÓS-PRODUÇÃO			
	Roteiro	Indentidade Visual	Vinheta	Planos	Idioma	Cenário	Iluminação	Áudio	Intervenções	Trilha Sonora	Cores	Fontes
SIMILARES DIRETOS												
Weder A./2013	Não	Não	Não	Geral	Português	Academia	Ambiente	Não	Não	Música pop	Não	Abertura e encerramento
Alice Pinho/ 2012	Talvez	Não	Não	Geral	Inglês	Clínica ou Academia	Ambiente	Não	Título de exercícios	Música calma	Não	Durante o vídeo
Revista da Cidade/2012	Sim	Sim	Não	Todos	Português	Estúdio	Estúdio	Sim, qualidade de estúdio	Título da matéria	Música ambiente	Tons frios	No título
Neurologia 21/2013	Sim	Não	Sim	Todos	Inglês e Espanhol	Sala	Ambiente e Estúdio	Sim, qualidade de estúdio, porém duplo	Não	Música ambiente	Tons quentes	Vinheta
Daiane Seibert/2014	Talvez	Não	Não	Geral e Médio	Português	Ambiente Externo	Ambiente	Sim, com ruído e trilha sonora se sobrepondo	Título de exercícios	Música pop	Não	Na abertura e durante o vídeo
Caio Vicentini/2012	Talvez	Não	Sim	Geral	Português	Casa da Paciente	Ambiente	Sim, com trilha sonora se sobrepondo	Não	Instrumental	Não	Em quadros de transição
Barbara C./2013	Não	Não	Não	Todos	Português	Casa da Paciente	Ambiente	Não	Não	Música eletrônica	Não	Abertura e encerramento
1 Parkinson/2009	Talvez	Não	Não	Geral	Inglês	Clínica ou Academia	Ambiente	Não	Não	Instrumental	Não	Em quadros de transição
Jean Paul/2011	Sim	Não	Não	Todos	Francês	Clínica e Casa	Ambiente	Sim, qualidade de estúdio	Título de exercícios	Instrumental	Não	Durante o vídeo
Jhon Argue/2011	Não	Não	Não	Geral	Inglês	Sala	Ambiente	Sim, qualidade de estúdio	Não	Instrumental	Não	Abertura e encerramento
CaptainPat/2011	Talvez	Não	Não	Todos	Inglês	Academia	Ambiente	Sim, qualidade de estúdio	Não	Instrumental	Tons quentes	Abertura e encerramento
SIMILARES INDIRETOS												
Home Fit Home/2015	Sim	Sim	Não	Todos	Português	Sala	Ambiente	Sim, qualidade de estúdio	Não	Não	Não	Não
Fernando G./2014	Sim	Sim	Não	Todos	Inglês	Estúdio	Artificial	Sim, com trilha sonora se sobrepondo	Não	Música agitada	Tons frios	Não
BeFit/2012	Sim	Sim	Não	Todos	Inglês	Estúdio	Ambiente	Sim, com ruído	Não	Música ambiente	Tons quentes	Não
Exercicioemcasa /2014	Sim	Sim	Não	Geral	Português	Sem Cenário	Artificial	Sim, qualidade de estúdio	Não	Música agitada	Não	Na abertura
NamuFit/2015	Sim	Sim	Sim	Todos	Português	Sala	Ambiente	Sim, qualidade de estúdio	Sim	Música ambiente	Tons frios	Na abertura e durante o vídeo

SOURCE: PREPARED BY THE AUTHORS BASED ON ANALYSIS OF SIMILARITIES

Based on an analysis of visual criteria such as capture quality, audio quality, color editing, visual identity, the presence of a vignette, among other requirements necessary for a quality production; according to bibliographical research carried out; it can be concluded that the majority of similar productions are not concerned with stages such as pre-production and post-production.

The examples analyzed have no vignette or visual identity, important elements for promoting and publicizing the work, as well as reinforcing the reliability of the content. As for editing, this is also absent, as there are videos with a homemade or amateurish image capture look.

It is also worth noting that exercise for Parkinson's sufferers is not a common topic, since among the direct similarities the newest video is from 2014.

The project emphasizes the importance of the presence and/or participation of a physiotherapy professional so as not to pose a risk to the patient performing the exercises. Most of the similar ones feature professional physical educators, with methods that differ from the treatment offered by a physiotherapist.

Thus, factors that were considered essential for the development of an audiovisual production suitable for Parkinson's patients are not taken into account in the production of direct similarities, and few examples of indirect similarities come close to meeting the parameters used for the analysis.

Chapter 3

3. CONCEPT

3.1 AUTONOMY AND RELAXATION

Contact with research and with people with Parkinson's disease has shown that a value that is very dear to patients is their autonomy. Independence, the ability to do one's own thing, is what in many cases strengthens the patient. As Nazari said in conversation, small everyday actions, when carried out without help, ease the psychological effects of the disease. Research has also shown that physiotherapy helps treat the disease and mitigates its effects. As such, the project will provide autonomy by being made available in media accessible to both groups of target audiences. The aim is also to convey this through exercises that are compatible with a patient's difficulties. The role of design will be to make the information clear and objective, so that patients can do their exercises without depending on third parties. The audiovisual production will have language appropriate to its audience, who according to research have vision and hearing problems, so the layout will be simple, without interference that could affect the perception of the information, in addition to the audio quality, which should be noise-free. In addition to the psychological and emotional well-being of Parkinson's sufferers, the production will complement the concept with relaxation. The analysis of similarities shows the use of neutral aspects or cold colors, which creates an environment reminiscent of a hospital, and brings a sense of sadness to the viewer. Complementary colors will be used, reinforcing the contrast between warm and cold colors to break up this feeling, but without losing the seriousness of the content.

Chapter 4

4. LIST OF REQUIREMENTS

Develop a visual identity for the project

Drawing up a roadmap with professionals Nazari and Ronchi

Define storyboard based on the script drawn up together with physiotherapy professionals

Produce a vignette that reinforces the visual identity and contains information relevant to the public, such as: the degree of illness appropriate for the exercise (based on a medical table), the time needed to perform it, what is being worked on with this exercise, etc...

Work on a dynamic recording, with interleaving of shots.

Capture studio-quality audio without noise.

Use a soundtrack suitable for the audience and consistent with the exercise, which does not interfere with the audio

Studio lighting in the production to reinforce the credibility of the videos

Presence or assistance of a physiotherapy professional to record the exercises

Graphic interventions, if appropriate, that do not interfere with the understanding of the image, and that are clear.

When it is necessary to use text, consider the vision problems of most users and use larger, legible type.

Colors: Work with complementary colors, reinforcing the contrast between cold and warm colors, both in the production and post-production phases, because according to an analysis of similar videos, it was noted that most of them work with cold colors, which are reminiscent of a hospital and make the production look sad.

In addition to being published on the association's YouTube and Facebook pages, the videos should also be recorded on DVD, as according to research this is a need of the target audience.

Chapter 5

5. DEVELOPMENT

5.1 PRE-PRODUCTION

In this pre-production stage, the *naming* will be chosen, the brand will be created, the color palette, graphic elements and typography will be defined. The videos will also be programmed, user preference tested, scripted and *storyboarded.*

5.1.1 *Naming*

To choose the name, we *brainstormed* some words that were directly linked to the project, such as: design, dopamine, exercise, physiotherapy, Parkinson's, therapy, tulip, video. Other words that were not directly linked to the project were also used, such as: self, well, walk, happy, flourish, strength, independence, revive, health, safety, sow, life, vibrate. Based on these words, ideas began to emerge for the project's nomenclature, such as: Autotherapy, learning with Parkinson's, well, flourish, walk happily, flourish, revive with Parkinson's, live therapy and fly. At first, we wanted the name to have several meanings and not say exactly what the project was about. That's why we looked for words that had meanings that were indirectly linked to the disease, as was the case with "Voe", which conveyed freedom, independence, which is one of the aims of the project, the independence of the sufferer. The first tests with the brand were using the name "Voe" and for this we used some symbols such as a bird and also a weather vane, so that they brought movement into the brand. However, during the development of this project, it was noticed that the nomenclature "voe" was inadequate, and that because it was a new and different project, it was necessary for its nomenclature to explain itself, in a more formal and serious way, so new studies were carried out in search of new ideas, where the name "EXPA" was arrived at, which means "exercises for Parkinson's", this idea came about because of the NAMU FIT Channel, where "NAMU" means "in change". As well as being the combination of the words exercise and Parkinson's, this name was also chosen because of its sound, which is similar to the word SPA.

FIGURE 65 - BRAINSTORM

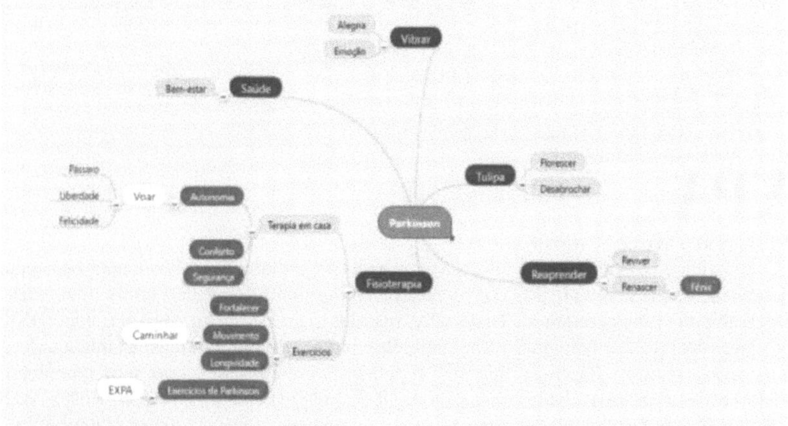

SOURCE: PREPARED BY THE AUTHORS

5.1.2 Brand

As previously mentioned, the idea for the brand was to have a *naming* and a symbol that indirectly referred to the project and that didn't depend on each other. At first, the main objective was to break away from the conventional and do something different from the main symbol of Parkinson's, which is the tulip flower. Before starting on the designs, a panel of references was made Fig.66, with designs that could break away from the common sense of the tulip and convey movement, one of the main factors of this project. As a result, alternatives emerged, such as the wind vane, the dopamine formula, birds, ideas that showed movement Fig.67.

FIGURE 66 - REFERENCE PANEL

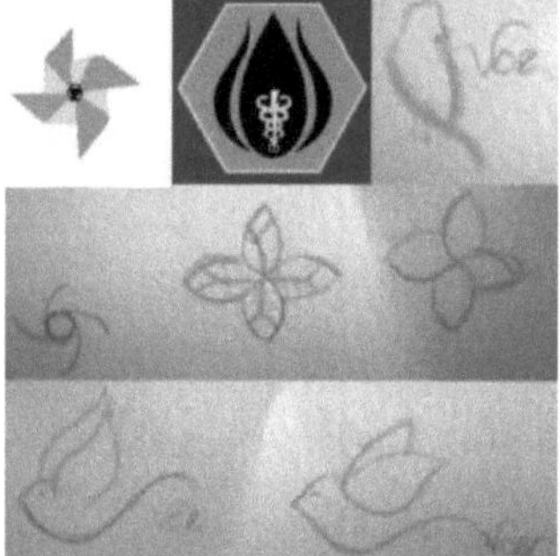

SOURCE: PREPARED BY THE AUTHORS
FIGURE 67 - ALTERNATIVES GENERATED

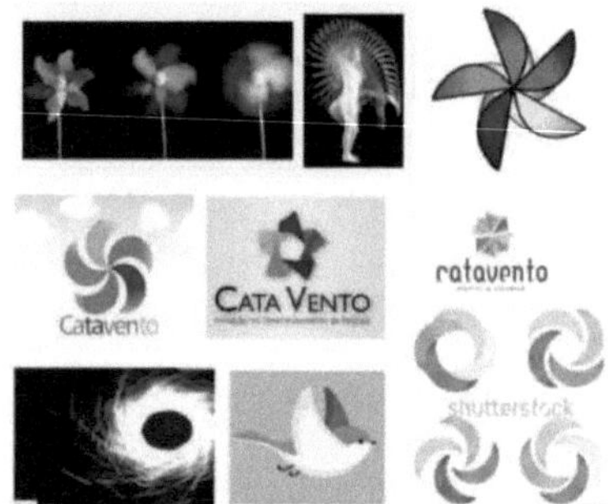

SOURCE: PREPARED BY THE AUTHORS
FIGURE 68 - TYPOGRAPHY

EXPA

SOURCE: PREPARED BY THE AUTHORS
FIGURE 69 - MERMAID TYPOGRAPHY

EXPA

SOURCE: PREPARED BY THE AUTHORS

But with the emergence of the *naming,* it was realized that it would no longer be necessary to have a symbol, as the name already conveyed everything that was needed on its own, such as the seriousness that the project required. That's why, in order to create the brand, it was concluded that it was only necessary to use typography and geometric shapes or simpler lines, as this would convey regularity and continuity. Based on this, two typefaces were chosen: a sans serif typeface Fig.68, because it is cleaner, more readable and has the right characteristics for the project, such as its regular lines, and a serif typeface Fig.69, because it is more classic and conveys a sense of continuity.

After defining the two sources that would be tested, the generation of alternatives began. The main objective was to maintain the seriousness and regularity of the project, as well as something cleaner, without too much noise and that could be seen clearly from a longer distance, mainly because the target audience has some difficulty seeing, so geometric shapes and lines were used Fig. 70.

FIGURE 70 - BRAND ALTERNATIVE GENERATION

After a few generations of alternatives, the ones chosen for the preference test with the target audience were those that best conveyed the idea of security and credibility, that had the least possible noise and good readability, one through a geometric shape and the other through a continuous line, both of which represent regularity and precision Fig.71.

FIGURE 71 - ALTERNATIVES CHOSEN FOR TESTING

5.1.3 Color palette

To define the color palette, it was necessary to work with warm colors and cool colors that complemented each other and could convey the joy and seriousness needed in this project. The goal with the color palette was to convey more joy to the patients and to bring something that did not at any time refer to the hospital or clinics, as these places themselves tend to be sadder and more sober, and their colors are also lighter and convey more sadness. That's why the main idea was to bring color, relaxation, but still be able to convey seriousness and credibility to the project, so that it wouldn't be too childish, since the target audience is older people who generally prefer more neutral colors.

It was with the target audience in mind that we came up with two palette options, one with brighter colors and the other with slightly more opaque colors, which will be chosen by the target audience through an online preference test Fig. 72 and 73. In addition to the colors having the objective of bringing more joy and relaxation to the project, their other function will be to divide the videos into categories, because for each category there will be a color, which will already be present from the vignette and also in the graphic elements that will be in the middle of the video, so it will make it easier for the user to better distinguish the categories.

FIGURES 72 AND 73 - SELECTED COLOR PALETTES

5.1.4 Typography

One of the typefaces chosen to be part of the graphic interventions during the video is already part of the brand's institutional alphabet. This is Quicksand, a sans serif typeface that is clean and allows for good readability and legitimacy Fig.74. The other typeface has a serif and is a little thicker, which is Mermaid Fig. 75. Both were presented in the preference test so that the user could choose the one that best suited them.

FIGURE 75 - QUICKSAND ALPHABET

ABCDEFGHIJkLMNOPQRSTUVWXYZ
abcdefghijklmnopqrstuvwxyz
0123456789
. , : ; ' " ( ) ?! + - */ =

SOURCE: PREPARED BY THE AUTHORS
FIGURE 75 - MERMAID ALPHABET

ABCDEFGHIJKLMNOPQRSTUVWXYZ
abcdefghijklmnopqrstuvwxyz
0123456789
. , : ; ' " ( ) ?! + - */ =

SOURCE: PREPARED BY THE AUTHORS

5.1.5 Graphic elements

The graphic elements that will be used are simple, following the same idea used when choosing the typography, which was cleaner type without noise. This project will only require direction icons such as arrows to better guide the user in the direction and movement being made.

FIGURE 76 - GRAPHIC ELEMENTS

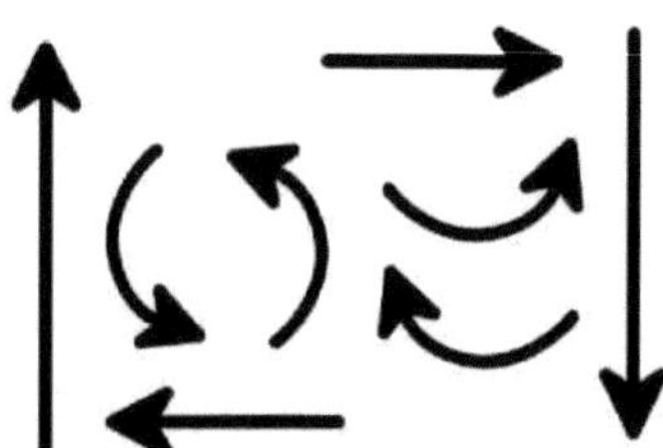

SOURCE: PREPARED BY THE AUTHORS

5.1.6 Video program

At this stage, together with the professional, it was decided that it would be best to separate the videos into five categories, such as: introduction to the channel, where the physiotherapist will introduce herself and describe the objectives of the channel, speech therapy exercises, lying exercises, standing exercises and sitting exercises. This is so that there is no risk to the patient, because if they had to move too much they could easily lose their balance and fall. These categories will be described both in the title of the video and in an information screen that appears at the beginning of the video and in its description. They will also be divided according to the colors of the palette, each color representing a specific category that will be used in the entire video, including the vignette and visual identity.

5.1.7 Preference test

To check user preferences, an online questionnaire was carried out with 20 people, with three questions relating to the brand's typography, branding and color palette.

When asked about the source of the project, 60% of people chose source 1 and 40% chose source 2, as can be seen in the graph below:

GRAPH 7 - RESPONSE TO THE ONLINE QUESTIONNAIRE

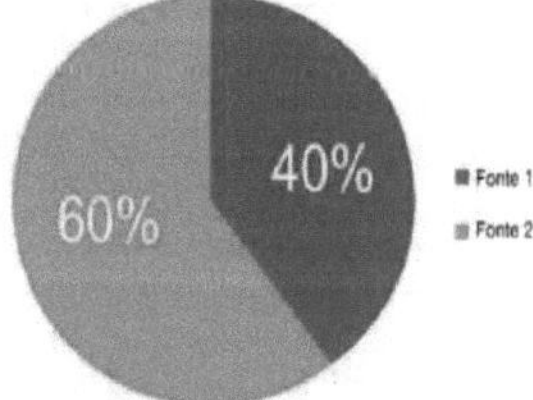

SOURCE: PREPARED BY THE AUTHORS

The second question asked about the preferred color palette and 85% of users opted for palette

1.

In the graph, you can see that 60% of the public opted for font 2, which in this case is Quicksand Book, which as I said before belongs to the brand's typographic family.

GRAPH 8 - RESPONSE TO THE ONLINE QUESTIONNAIRE

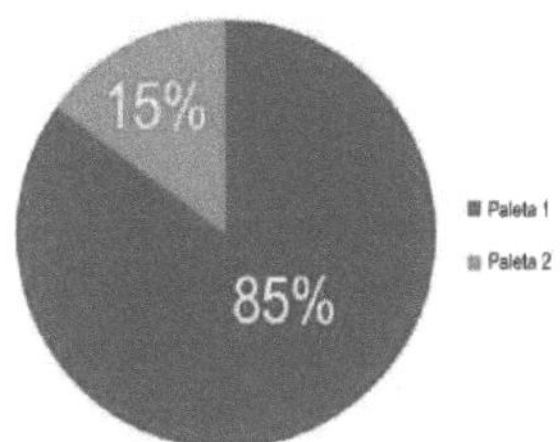

SOURCE: PREPARED BY THE AUTHORS

FIGURES 77 AND 78 - QUICKSAND FONT

SOURCE: PREPARED BY THE AUTHORS

Palette 1 has more vibrant and cheerful colors, as you can see below:

FIGURE 79 - COLOR PALETTE

SOURCE: PREPARED BY THE AUTHORS

And the last question was related to the brand, to find out which one users liked best and the result was that 70% liked option 1 best.

GRAPH 9 - RESPONSE TO THE ONLINE QUESTIONNAIRE

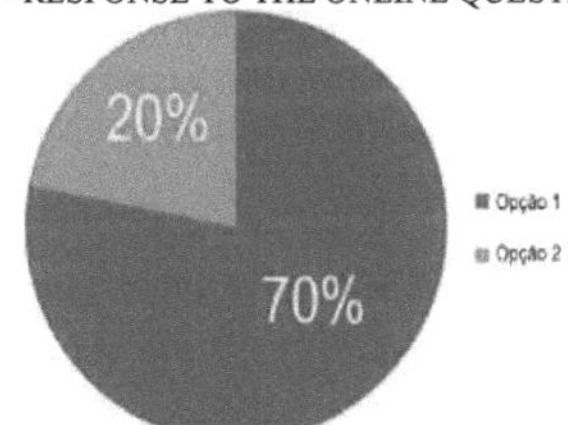

SOURCE: PREPARED BY THE AUTHORS

So the chosen option is:

FIGURE 80 - OPTION 1 CHOSEN

SOURCE: PREPARED BY THE AUTHORS

5.1.8 Roadmap

Based on the book Parkinson's Disease by the author Telmo Reis (2004), with adaptations by the physiotherapist Patricia Nazari, a sequence of exercises was defined, which were divided into four videos according to the position in which they are performed: sitting, lying down, standing and facial mimicry, which will also be performed in the sitting position, but with a different specialty.

FIGURE 81 - BOOK PARKINSON'S DISEASE BY TELMO REIS

SOURCE: http://www.neurologiacirurgica.com.br/artigos/detalhes/id/12

5.1.8.1 Vignette script

The opening vignette will start with the acronym EXPA which is the brand name, then there will be a click with the mouse cursor which will simulate a button, and then the acronym will expand, transforming into its meaning "Parkinson's Exercises".

To close the videos, the vignette will take the opposite route in its animation, starting with the meaning of the acronym, the click simulating a button will happen again, and the words will close, returning to the acronym EXPA.

5.1.8.2 General Roadmap

All the videos will start with the vignette followed by a screen with information on the position in which the exercises will be performed. When there are materials to be used, there will be a screen with a list of all the items.

The repetitions of the exercises will follow a pattern stipulated by the physiotherapist, this information will also be at the beginning of each video, on a screen after the materials screen.

Between exercises there will be a transition screen, announcing the number of the next exercise and which material will be used in it.

In the first exercise, as soon as the physiotherapist appears, in the bottom left-hand corner of the screen, there will be an animated rectangle with information such as the professional's name and profession.

The graphic interventions, as well as the verbal reinforcement, will take place during the videos as necessary. It was decided to outsource the animation, which will be carried out by Wagner Regis according to the authors' ideas.

At the end, the closing vignette will be used, followed by an invitation screen to subscribe to the channel and a support screen with the brands of Design, Positivo University and the Paraná Association of Parkinson's Disease Patients.

5.1.8.3 Video of seated exercises

In this video, 6 exercises will be performed.

The first exercise will be filmed in medium shot, as it will use a table for support and the exercise is a series that helps with hand control, especially for signatures. Materials needed: chair, table and pen.

The second exercise will be filmed first in a frontal shot and then in a lateral shot. It's an exercise that needs this shot and lateral reinforcement to show the posture and inclination. Materials needed: chair and broom.

The third exercise will be filmed in a lateral American shot, as it shows the initial sitting position and the final standing position, but the focus of the exercise is on the torso and not the feet. Materials needed: chair and table.

The fourth exercise will be filmed in medium shot, as the exercise will emphasize the movement of the head, shoulder, torso and arms. Materials needed: chair.

The fifth exercise will be filmed in an American shot. It will work on the same areas as the previous exercise, but with the torso leaning forward, this shot will demonstrate this movement in a more understandable way. Materials needed: chair.

The sixth exercise will be filmed in the foreground, it is an exercise that needs to focus on a specific area, the feet. Materials needed: chair, ball or similar.

5.1.8.4 Facial mime video

In this video, 10 exercises will be performed.

All the exercises will be filmed in medium shot, as they need to focus on the face and torso area. The first five exercises will work on breathing, stretching and moving the neck area. And the last five will be facial mime. No materials required.

5.1.8.5 Video of standing exercises

In this video we will perform 07 exercises.

This video will be filmed in close-up to demonstrate the exercise in its entirety, and in medium shot for the description of each one.

The first three videos will use a chair for support as they work on leg movements which require balance. The fourth exercise will be unsupported due to space, and the possibility of wall support will be indicated in the edit. The fifth and sixth exercises will work on arm stretching and posture, requiring a large ball (soccer, volleyball, basketball) and a broom. The last exercise will work on stride (walking).

5.1.8.6 Lying down exercise video

In this video, 10 exercises will be performed.

This video will be filmed in medium shot for the descriptions of the exercises, lateral general shot and superior general shot for the exercises, according to the best representation for each one.

Materials needed: All the exercises will require a bed, mat, mattress or sofa (surface to lie on).

Exercise 1, 2, 3, 4 and 5 - No material required.

Exercise 6 - Broom.

Exercise 7 - Towel.

Exercise 8 and 9 - Big ball (soccer, volleyball, basketball...).

Exercise 10 - Tennis ball or similar.

5.1.8.7 Channel presentation video

In this video, physiotherapist Patricia Nazari will introduce herself, talk about her experience, give a brief description of the channel and invite people to subscribe to follow the videos. Her speech will be accompanied by the vignette soundtrack.

The opening and closing vignettes will be used. When she talks about the channel, a link to subscribe will appear at the bottom of the screen.

In the DVD version, the part about signing up will be cut.

5.1.8.9 Animações script

Video 1: Sitting position

-The arrows must be colored #edbc32

Exercise 1:

A spring-like arrow enters from the physiotherapist's right and stretches with the movement of the hand.

In rotational movements, one arrow goes in and the other goes behind it. One rotates behind the other, according to the movement of the hand in both directions.

The arrow flexes as the physiotherapist moves it.

Exercise 2:

The arrow representing the posture will form an angle of inclination.

The arrow that will represent the movement of the arms with the broom enters from the right (behind the physiotherapist), and moves along with the movement, repeating the movement to the right.

Exercise 3:

The arrow rises as the physiotherapist stands up and falls as she sits down.

Exercise 4:

Arrow enters from the right and curves according to movement, repeats from the left.

Exercise 5:

Two arrows enter, one from the right and one from the left, pushing the physiotherapist's arms. The arrow enters at the top of the screen and tilts in a curve, indicating the shrinking of the torso during the exercise.

Exercise 6:

Arrows that follow the movements of the feet, forwards and backwards in circles.

Video 2: Seated pose - Facial mime video

-The arrows must be in color #B72943

Exercise 1:

Arrow with inclination (bulging downwards) pushes the physiotherapist's head, following the exercise one side at a time.

Exercise 2:

Verbal reinforcement only.

Exercise 3:

Same situation as exercise one, but without the tilt. Straight arrows pushing the physiotherapist in the direction of the exercise.

Exercise 4:

The arrow enters from the top of the screen, and moves downwards to both sides following the movement. (Something like the animation in exercise 4 of the first video.)

Exercise 5:

Two arrows enter simultaneously from both sides, coming from the bottom of the video following the shoulders upwards, and descend together with the physiotherapist's arms.

Exercise 6:

Verbal reinforcement only.

Exercise 7:

Verbal reinforcement only.

Exercise 8:

Arrow to both sides. It can be on the left side of the video, following the physiotherapist's tongue.

Exercise 9:

Verbal reinforcement only.

Exercise 10:

Verbal reinforcement only.

Video 3: Lying position

-The arrows must be colored #0097bb

Exercise 1:

Arrow on the spine, just to indicate posture. Arrow on leg accompanying movement, when she touches the first leg to the mattress arrow disappears, another appears on the other leg, disappears when she touches the mattress. When she lifts both legs, two arrows follow the movement.

Exercise 2:

Verbal reinforcement only.

Exercise 3:

Arrow on the leg as in exercise 1, following the movement, first on one and then on the other. Arrow to indicate the movement of the arm: First there's a left arrow tilted to the right at the top of the screen, at the second point in the exercise the arrow comes from the bottom.

Exercise 4:

Arrow bent as hips and thighs stretch with the movement.

Exercise 5:

Arrow on top of the leg following the movement.

Exercise 6:

Arrow pointing upwards, indicating the position of the arms. Another arrow above the leg, accompanying the movement performed.

Exercise 7:

Straight arrow on top of the leg, pointing upwards. When you go down, this arrow disappears, and appears when the other leg is stretched out.

Exercise 8:

The arrow moves like a pointer, above the arm, following the movement.

Exercise 9:

Seta enters from the right and curves upwards as if pushing the physiotherapist's head.

Exercise 10:

The arrow follows the movement of the ball.

Video 4: Standing position

-The arrows must be colored # f47b41

Exercise 1:

Arrow does as in the last exercise in the video in a sitting position, but when she bends her leg, the arrow bends along with it.

Exercise 2:

Arrow as in exercise 1, only vertical, from top to bottom, inclined to the left.

Exercise 3:

Arrow following foot movements, leaning forwards and backwards.

Exercise 4:

An arrow next to the physiotherapist's body, indicating the same angle of bending of the leg, and another arrow coming from the bottom with an inclination to the right and upwards. The second part keeps the angle arrow, and the other arrow comes from the top of the screen with a right and downward tilt.

Exercise 5:

Seta pushes the physiotherapist from the left side to the right, passes her and pushes from the other side.

Exercise 6:

Spring-shaped arrow pointing upwards, when the physiotherapist lowers the broom the spring tightens, and releases when she raises the broom.

Exercise 7:

Verbal reinforcement only.

5.1.9 *Storyboard*

Below are images of the *storyboards* developed, detailing each of the channel's videos, as well as its vignette.

FIGURE 82: STORYBOARD OF THE VIGNETTE

SOURCE: PREPARED BY THE AUTHORS

FIGURE 83 - STORYBOARD OF THE PRESENTATION

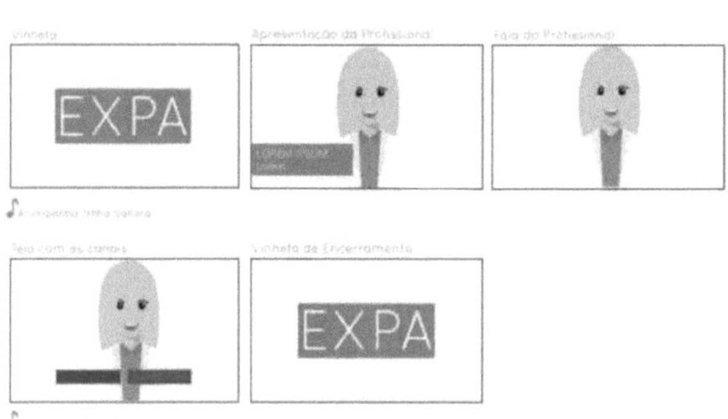

SOURCE: PREPARED BY THE AUTHORS

FIGURE 84- STORYBOARD SITTING POSITION

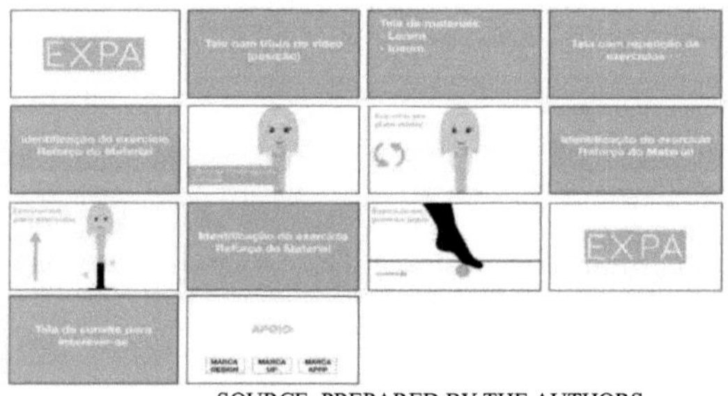

SOURCE: PREPARED BY THE AUTHORS

FIGURE 85 - STORYBOARD **FACIAL** MIME

SOURCE: PREPARED BY THE AUTHORS

FIGURE 86 - STORYBOARD STANDING POSITION

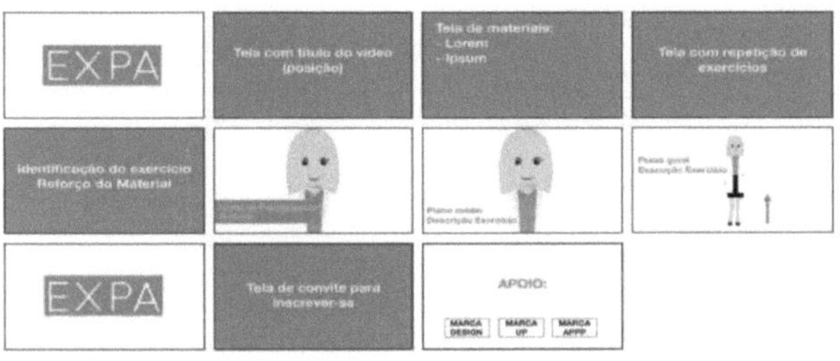

SOURCE: PREPARED BY THE AUTHORS
FIGURE 87 - STORYBOARD LYING POSITION

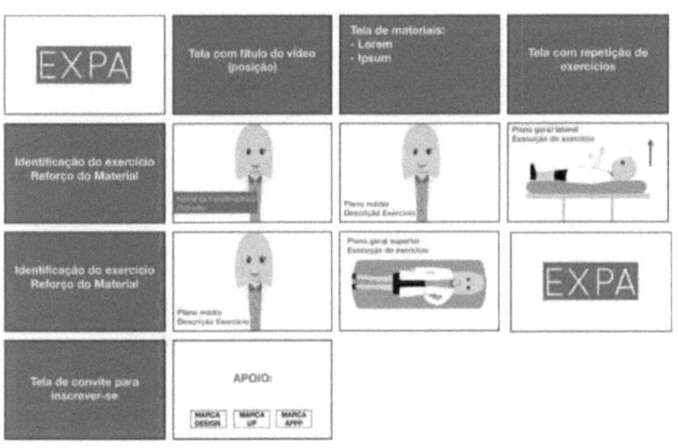

SOURCE: PREPARED BY THE AUTHORS

In general, the videos follow the pattern described above in the storyboard, with a vignette and information screens. In the storyboard, it is planned which videos will not have certain screens, and which shots will be used in each one, as well as the color programming. The shots were chosen to make it easier for the user to understand each exercise.

The images represent the information screens, and the filming plans are visual representations of the planning based on the script. Changes may occur after validation.

5.2 PRODUCTION

At the production stage, before recording began, the script was sent to physiotherapist Patricia Nazari, with the order in which the exercises would be performed. Before recording began, light tests, camera tests, framing tests, set tests and microphone tests were carried out. It took three days to make the recordings, one for the tests and two for the videos. For the videos, it was necessary to take some materials that would be used to do the exercises. We decided on materials that are easily found in any home, such as a broom, a ball, a 1kg packet of food, a small ball or an orange and a towel. The technical materials used were a Nikon D7000 camera, two Qi torches for lighting, a 2002 *mako* flash and a tripod to capture the images, while the *loud wm-1001* microphone was used to capture the sound. Below are photos of the *making off*:

FIGURES 88,89,90 AND 91 - MAKING OFF

SOURCE: PREPARED BY THE AUTHORS

5.2.1 WINE

In this project, the vignette aims to reinforce the brand and explain its meaning. To do this, we thought of something simple that follows the identity of the project. The production was done in Adobe Photoshop CC and Adobe Premiere CC. It is an animation of the brand, where it is first presented with the geometric element (the rectangle), four versions were developed, each version will have a different color, the colors of the palette. Under the rectangle appears the brand's typography, then the cursor of a mouse that clicks on the rectangle, assimilating it with a button, the brand shrinks making the effect of a button when clicked, and then extends to the sides revealing the full name "Parkinson's Exercises", and so it ends. The mouse cursor was animated in .png format frame by frame in Adobe Photoshop, then inserted into the Adobe Premiere project, where the rest of the animation was developed in frames. The entire animation is accompanied by the "SunshineVer   " soundtrack, which is available on the Incompetech website   at http://incompetech.com/music/royalty- free/index.html?feels%5B%5D=Relaxed&page=13

FIGURES 92, 93, 94, 95 and 96 - WINE ANIMATION SEQUENCE

SOURCE: PREPARED BY THE AUTHORS

5.3 POST-PRODUCTION

After production, we started editing the videos and adding graphic elements to the project. As previously decided with the professional, the videos were divided into categories: exercises practiced in the sitting position, standing, lying down and speech therapy exercises. To indicate which category each video will be in, a screen with the category title appears after the vignette. This is followed by a new screen in the same color, where all the materials that will be used during the video are listed, and then the number of repetitions of the exercises in that video. The next screen is a transition of scenes and reinforcement of material, following the pattern of the other two screens, where the typography appears this time with the number of the exercise and what material will be needed to perform it. This screen will be used every time the exercise changes. All screens follow the color selected in the palette for that category, and the typeface chosen for interventions, Helvetica. In the first exercise of each video, an intervention was also inserted to introduce the professional, which is a rectangle that slides at the bottom left and when it stabilizes, Patricia's name and profession appear.

FIGURE 97 - TRANSITION SCREEN

SOURCE: PREPARED BY THE AUTHORS
FIGURE 98 - PRESENTATION

SOURCE: PREPARED BY THE AUTHORS
FIGURE 99 - INFORMATION SCREEN

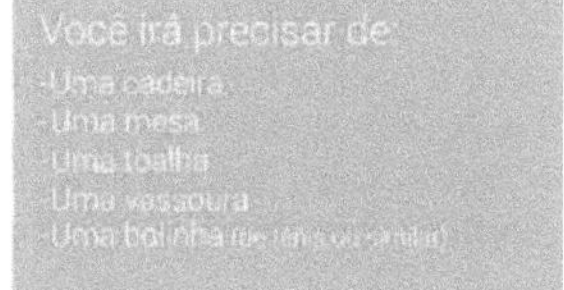

SOURCE: PREPARED BY THE AUTHORS

FIGURE 100 - REPEAT SCREEN

SOURCE: PREPARED BY THE AUTHORS
FIGURE 101 - CATEGORY

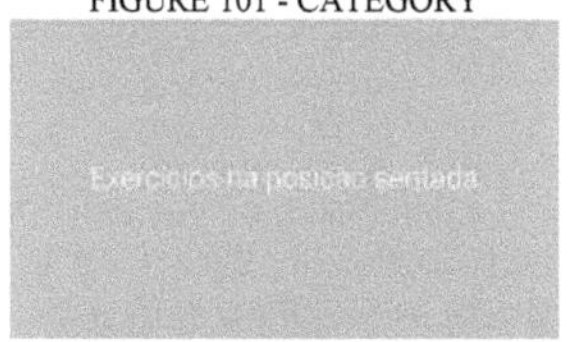

SOURCE: PREPARED BY THE AUTHORS

The videos underwent light and color treatment and some cuts were made to each exercise. It was decided to use a dry cut so as not to confuse the audience, followed by the transition screen mentioned above.

FIGURE 102 - ORIGINAL VIDEO

SOURCE: PREPARED BY THE AUTHORS
FIGURE 103 - VIDEO WITH EDITING

SOURCE: PREPARED BY THE AUTHORS

The audio, which was captured using a lavalier microphone and recorded with Adobe Audition, was edited to eliminate noise in the same program, then synchronized in Adobe Premiere.

Graphic interventions were then inserted to reinforce the movement. At first they were made in Adobe Illustrator, then transferred to Adobe Photoshop, saved in .png file format, and then imported into the project in Adobe Premiere, where they were animated in movements that accompanied the exercises, their colors chosen according to the color palette used for each video. The decision was then made to outsource the animations so that the quality of the videos could be maintained. The authors scripted the animations and the teacher and animator Wagner Regis developed them. According to audience research, there was a need for information to be reinforced in various cases of difficulty. For this reason, in addition to the graphic interventions, which are a pictorial reinforcement of the project, words were inserted to reaffirm the project's success.
the movement, thus being a verbal reinforcement, both of which complement the visual and audio information.

FIGURES 104, 105 AND 106 - GRAPHIC INTERVENTIONS

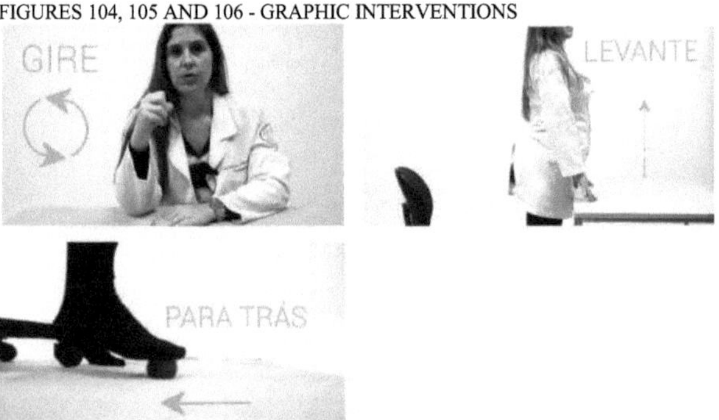

SOURCE: PREPARED BY THE AUTHORS

To end the video, there is a screen inviting the viewer to subscribe to the channel to follow the next videos, made in the same pattern as the initial information screens. The reverse vignette has been inserted, and the closing screen features the brands of Positivo University, the Design sector and APPP.

FIGURE 107 - INVITATION SCREEN

SOURCE: PREPARED BY THE AUTHORS
FIGURE 108 - SUPPORT SCREEN

SOURCE: PREPARED BY THE AUTHORS

5.4 VALIDATION

As a form of validation, we opted for a qualitative analysis through a focus group.

A focus group is a form of validation where a small group of people get together to evaluate a particular product by using it. This group is observed while using the product, and at the end a conversation is held where a moderator encourages the group to talk about certain evaluation points.

For the project, the validation took place at the APPP, with four patients and coordinator Vanessa Szuba, with the authors and physiotherapist Patricia Nazari as moderators. The video was played in a seated position, with the four patients sitting in front of the computer, with a table in front of them with the necessary materials available. The physiotherapist and the coordinator observed the performance, while the authors documented it with photos and notes.

During the performance, it was noted that there were no difficulties in interpreting the exercises, only personal restrictions in one of the patients. At the end of the reproduction, the patients had a brief moment to discuss the activity with each other. We heard the coordinator's comments, which were about the length of the information on the screen, which according to her should be longer, and about the exercise repetition information, which until then appeared on a screen at the beginning as standard. She noticed that this way the patients didn't repeat the exercises and followed the sequence of the video. Vanessa suggested that at the end of each video there should be a screen asking patients to pause the video and repeat the exercise. In conversation with the patients, there was the same concern about the length of the information, and one of them pointed out that he would like to see audio reinforcement on screens where only text appears, as on the materials screen. They pointed out that the information during the video was easy to understand, and that they really liked this option to reinforce physiotherapy. They even commented on the possibility of getting together at one of their homes to do the exercises, thus motivating each other.

With this validation, it can be concluded that changes will be necessary in order to make better use of the production for the target audience, and details that could only be perceived during execution will be refined.

FIGURES 109 AND 110 - VALIDATION WITH THE USER

SOURCE: PREPARED BY THE AUTHORS

5.5 BUSINESS PLAN

In order to find out whether this project would be viable or unviable for the authors and their target audience, it was necessary to carry out a value survey. The following figures were found:
- R$ 300.00 for recording the videos;
- R$ 400.00 for editing and animating the videos.

And since the main objective of this project is to share it on social networks, the idea would be to share it with several people so that everyone has access to this information, and so that there are more views and also more subscribers, so that through some advertisements broadcast on YouTube it is possible to make this project viable.

Another possible future for this project is, together with physiotherapist Patricia Nazari, to create a booklet containing the same exercises, to be distributed to patients who are unable to watch the videos, so that they can learn them from pictures.

FINAL CONSIDERATIONS

According to the results obtained through validation, the public was receptive to the project and the professionals involved showed great enthusiasm. There are still some adjustments to be made, and it will be necessary to find more resources if the project is to continue. The aim of the project began as a way of promoting autonomy and relaxation, but the validation revealed new possibilities. In the research carried out at the start of the project, it was pointed out that due to the limitations of the disease, many patients become depressed or less sociable. With the suggestion made by one of the patients who took part in the validation, about gathering friends at home to do EXPA exercises, we realized that it could help with this other aspect of the disease. We are therefore thinking of the possibility of videos with exercises in small groups or pairs, but without abandoning the initial idea of videos that can be done independently. A variation for the growth of the channel and a wider audience.

ALVES, M. N.; ANTONIUTTI, C. L.; FONTOURA, M. **Mídia e produgäo audiovisual**: Uma introdujo. 20 ed. Curitiba: IBPEX, 2008.
BURGESS, J.; GREEN, J **Youtube and the digital revolution**: How the biggest phenomenon in participatory culture has transformed media and society. 1 ed. Sao Paulo: Aleph, 2009.
COMPARATO, Doc. **Da criagäo ao roteiro**. 5 ed. RIO DE JANEIRO: ROCCO, 1995.
FERRAZ, Henrique Ballalai; BORGES, Vanderci. **How to diagnose and treat Parkinson's disease.** Discipline of Neurology, Federal University of Sao Paulo - Paulista School of Medicine (Unifesp - EPM). Available at: http://www.moreirajr.com.br/revistas.asp?fase=r003&id_materia=1870. Accessed on June 14, 2016.
GARBIN, H. B. R.; NETO, A. F.P.; GUILAM, M. C. R. The internet, the expert patient and medical practice: a bibliographical analysis. **Interface**, v. 12, n. 26, September 2008.
GROSSMANN, Marylandes. **With Parkinson's and doing well with life: the experience of a parkinsonian.** Sao Paulo: Lemos Editorial, 1998.
LANA, R.C.; ÁLVARES, L.M.R.S.; NASCIUTTI-PRUDENTE, C; GOULART, F.R.P; TEXEIRA-SAMELA, L.F.; CARDOSO, FE. **Perception of the Quality of Life of Individuals with Parkinson's Disease through the PDQ-39.** Brazilian Journal of Physiotherapy. Available at: http://www.scielo.br/pdf/rbfis/v11n5/a11v11n5. Accessed on: 05 Apr. 2016.
LIMA, Fabiola Mariana R. **Parkinson's disease.** Faculty of Medical Sciences of Paraiba . Available at: http://www.cienciasmedicas.com.br/artigos/2008/10/22/doenca-de-parkinson. Accessed June 14, 2016.
MARTIN, M. **A linguagem cinematográfica**. 4 ed. Sao Paulo: Brasiliense, 2004.
MCCLOUD, Scott. **Drawing comics**. 1 ed. Sao Paulo: M. BOOKS, 2008.
NAZARI, P.M.B Physiotherapist at the Paraná Association of Parkinson's Disease Patients. **Interview with Thais Santos and Thayza Paiva, held at the Association.** Curitiba, April 2016.
REIS, Telmo. **Parkinson's disease: patients, families and caregivers.** Porto Alegre: Pallotti, 2004.
RONCHI, D. C. M. Physiotherapist and Master of the Physiotherapy course at Positivo University. **Interview given to Thais Santos and Thayza Paiva, held at Positivo University.** Curitiba, April 2016.
SANT, Cíntia Ribeiro; OLIVEIRA, Sheila Gemelli; ROSA, Emerson Luis; SANDRI, Joice; DURANTE, Mirian; POSSER, Simone Regina. **Physiotherapeutic approach in Parkinson's disease.** Brazilian Journal of Aging Sciences Human. Available at: http://www.upf.br/seer/index.php/rbceh/article/view/259/194. Accessed on: June 14, 2016.
SANTOS, Viviane V.; LEITE, Marco Antonio A.; SILVEIRA, Renata; ANTONIOLLI, Reny; NASCIMENTO, Osvaldo JM.; FREITAS, Marcos RG. **Physiotherapy in Parkinson's Disease: A brief review.** Revista Brasileira de Neurología. Volume 46. Available at: http://files.bvs.br/upload/S/0101-8469/2010/v46n2/a0002.pdf. Accessed on: 05 Apr. 2016.
TELLES, André. **Digital Generation**: How to plan your marketing for a generation that searches on Google, connects on Orkut, sends messages by cell phone, opines on Blogs, communicates on MSN and watches videos on YouTube. 1 ed. Sao Paulo: Landscape, 2009.
THINK WITII GOOGLE. **Survey reveals Brazilians' intimacy with YouTube**. Available at:<https://www.thinkwithgoogle.com/intl/pt- br/articles/intimidade-dos-brasileiros-com youtube.html>. Accessed on: April 13, 2016.

APPENDICES

Interview given by Daiane Cristine Martins Ronchi, Physiotherapist and Master of Physiotherapy at Positivo University, specialist in Neurofunctional Physiotherapy, on April 28, 2016 at 9am.

Thais: Our goal with this interview is to search for content.

Daiane: So, tell me a bit about your project.

Thais: So the idea is to make a series of videos, there would be five, with physiotherapy exercises that patients can do at home. Something that doesn't pose a risk, so that they can continue the process done at the clinic.

Daiane: Cool, and your focus is design, right?

Thais: Yes!

Daiane: So your focus is the videos?

Thais: Yeah, we'd like to propose a better visual language, there are lots of videos on the internet, but none...

Daiane: Yeah...

Thais: (Laughs) Have you ever followed one?

Daiane: No, I don't look because there are so many bad things. We see things being done to patients that give us 10 kinds of despair! So, what do you need? **Thais**: So, we need bibliographical references.

Daiane: I'll email you some article references, because there's so little information in books. It's a disease that, despite being widely studied... Well, I'm going to look for more recent articles, so if you want help putting the text together... Because you don't even have the basis, right, so if you want, I have some students who are very interested in getting involved with these things, then we'll put it together. I need to know how many paragraphs or so you're going to dedicate to this, is yours a monograph or an article?

Thais: It's both.

Daiane: So it's good that you can put in more content, in physics it's an article. So they have a limit of twenty pages, including the cover and references.

Thais: In ours, the more things...

Daiane: So the body of the text is about ten pages long. So you can't cram too much in. So I see, when do you have to hand in this phase? **Thais:** In June, before the vacations.

Daiane: There's time. I'll get two students together who are really interested in this, and we'll do a pathology revision to give you a hand. But read on, read on. And if one day you want to make an appointment, I'll give you a more layman's explanation so you can understand how pathology works, even to make it easier for you when it comes to defending your work.

Thais: Yes.

Daiane: So that you can explain, in a way... Of course, the professors on the exam board aren't from the health field either, I don't know if you can call in people from outside, I don't know what it's like, but so that you can explain in a way that's more: "No, I understand the subject and I'm going to talk about it." So I'm going to look for some articles, the most recent and good ones, including articles that talk about the physiotherapy approach for Parkinson's patients, and then we can extract the exercises from the articles. You know, something with a theoretical basis, really cool.

Thais: We partnered with the Parkinson's association here...

Daiane: That's great!

Thais: And we'll talk to the patients there to find out what they need, we have to listen to them too.

Daiane: Yes, great! Did you do a questionnaire or something to talk to them?

Thais: We're going to elaborate this week, because we went there to really get to know them.

Daiane: If you want to do it and send it to me, so I can take a look, see if anything is missing, if anything needs to be included, send it to me! I'll do it, I'll help you do it. So there's no lack of information. If you're going to approach patients, make sure there's no lack of information. Because if you don't, then you'll have to keep going back to see... And another thing, this then goes into your work, right, where did you get it from? Why was it done this way? Yeah? In fact, there are some interesting articles that we can extract exercises from there, which are about Gametherapy... It's therapy with Wii, Xbox, that sort of thing. We can get a lot from there too.

Thais: Have you ever worked with the elderly?

Daiane: So, my specialty is neuropediatrics, but I've been supervising internships ever since I started teaching. Today at APAP we have three Parkinson's patients. So, last year I supervised an internship at the nursing home when the teacher in charge went on maternity leave and I took her place... I end up getting involved, I get involved well. And I'm always on the exam board in this area, because I specialize in adult and child neuro, but I focus on children, but I specialize in both.

Thais: But do you think they (the elderly) access the internet a lot?

Daiane: So today we have it easier. We have Seu Paulo who is a patient at APAP, and his wife is all connected, you know? And they're both doctors. You'll see, you'll deal with this public. So they have easy access. Another thing you could do is suddenly make a CD available.

Thais: That's the idea too.

Daiane: Because it's easier for them to put in a DVD and access it than to suddenly open it on Youtube.

Thais: We also want to make a DVD to give to patients.

Daiane: Yes, I think that for them, in terms of technology, access is easier, right? And so, sometimes the patient isn't, but the family caregiver is... The person who's involved, with the children, right, who look after the patient, they end up being.

Thais: Family is also a complicated issue, right, because many don't accept it, don't want to take care of it.

Daiane: Yes, it's a progressive disease. It only tends to get worse, there's no cure, treatment is palliative, there comes a time when it no longer has any effect. So, it really is a pathology, if you take a person with Parkinson's at the beginning of the disease, it's very difficult. It's not just the body that can't cope. It's more associated with pathology, it's a lack of neurotransmitters. So I'm going to separate the articles for you, so if you think it's too difficult for you to understand, because you don't really have the basis for it, right, to understand. Then we'll sit down sometime, even if we make an appointment, I have a break on Thursday from 10:20am until 11:10am, then one day we'll sit down and I'll explain it to you properly, okay, so that you can write the text knowing what you're talking about.

Thais: We need to prove it, right, where we're getting it from...

Daiane: Yes, that's why. You're going to have to convince people who aren't in the area that your work is necessary.

Thais: And see if they think he really is, right?

Daiane: I think, even if you want to continue afterwards... Is it the last year?

Thais: Yes.

Daiane: If you want to continue afterwards, next year we could see a TCC student who applies, you know, who makes a comparison before and after, for so many weeks using the program.

Thais: Yes, our idea is to continue. We want to post it on YouTube next year, on Parkinson's Day, post it on the association's Facebook page...

Daiane: Next year you'll take the fourth-year physio students, and then they'll put a scale to measure before and after exercise, of course!

Thais: A, that one, I never understood that scale.

Daiane: There are several, from the simplest to the most complex. Next year we can put them together, then you'll take part in the CBT. For I don't know, six weeks of a home program, with the videos, right, then we do a before and after and see if it's really having an effect. It's cool! But I'm going to send you the articles. I can't guarantee it'll be tomorrow because I have exams and an internship all day, but leave me your e-mail address. I've already got your e-mail address, so I'll collect about five for you to start with, read them, check them out and ask me anything, there will be words you'll read and you'll say: "Oh my God, what is this?". Okay, so if you google some definitions you'll find them, some will be easy to understand, others will make things worse.

Thais: Yes, some of them I'm already doing...

Daiane: It's just that it's a technical term, right, even physiotherapy students who are used to it have difficulty, you know? But that's how we do it, okay?

Thais: All right!

Daiane: So whatever you need to do, let me know. I'll sit down one day and explain the pathophysiology, how the disease happens, so that it's easier for you to work on the exercises, because they'll have to be very simple.

Thais: Yes, we'll have to, we agreed with the physiotherapist at the clinic, she'll do it from there.

Daiane: Great, if you'd like to pass on her contact details to me later, then we'll talk and do it together, she'll see her patient's needs, I'll see the needs of the ones I have and from experience, right, and then we'll put something simple into the patient's job that they can do, simply and in a short time, you know?

Thais: Yeah, they don't have much patience to keep up...

Daiane: So, if you think about the context at home, your routine at home, what would it be like to include a group of activities?

Thais: It's quite difficult, isn't it?

Daiane: So we have to think about it, right, there are patients who adhere, others who don't, but we'll measure that later. I'll pass it on to you, we'll talk.

Thais: All right then!

2.1 QUESTIONNAIRE

- How old are you?
- How old were you when you found out you had Parkinson's?
- Do you consider physiotherapy important as part of your treatment?
- Which type of exercise do you find most difficult? Why?
- Do you spend time at home practicing what you've learned in the clinic?
If so, what time of day and for how long?
- Do you usually use the internet?
- If you do, what are your activities on the computer? ( ) Watching videos ( ) Accessing news ( ) Social networks ( ) Other
- Do you have a DVD player?
- How would you like to access videos? ( ) Desktop computer ( ) Notebook
( ) Tablet ( ) Smart Tv ( ) DVD ( ) Other
- Do you have a family member who helps you with your day-to-day activities? If so, how old is this person?
- Do you have any health problems related to sight or hearing? If so, which ones?

ANNEXES

IMAGE USE AGREEMENT

TERMO DE AUTORIZAÇÃO DE USO DE IMAGEM E VOZ
Pessoa maior de 18 anos

Neste ato, e para todos os fins em direito admitidos, autorizo expressamente a utilização da minha imagem e voz, em caráter definitivo e gratuito, constante em fotos e filmagens decorrentes da minha participação no Trabalho de Conclusão do Curso TCC de Design de Projeto Visual da Universidade Positivo, a seguir discriminado:

Título do projeto: Produção Audiovisual : série de vídeos com exercícios para portadores de Parkinson com apoio do design.

Pesquisador (es) Thaís de Lima Santos e Thayan Karen de Paiva

Orientador(a): Gabrielli Harttmann Grimm

As imagens e a voz poderão ser exibidas: nos relatórios parcial e final do referido projeto, na apresentação áudio-visual do mesmo, em publicações e divulgações acadêmicas e nas redes sociais, em festivais e premiações nacionais e internacionais, assim como disponibilizadas no banco de imagens resultante da pesquisa e na Internet, fazendo-se constar os devidos créditos.

O aluno fica autorizado a executar a edição e montagem das fotos e filmagens, conduzindo as reproduções que entender necessárias, bem como a produzir os respectivos materiais de comunicação, respeitando sempre os fins aqui estipulados.

Por ser esta a expressão de minha vontade, nada terei a reclamar a título de direitos conexos a minha imagem e voz ou qualquer outro.

Curitiba, 25 de agosto de 2016.

Patricia M. Bariquello Nazari
Assinatura

Nome: PATRICIA M. BARIQUELLO NAZARI
RG: 7.222.267-9   CPF: 028.998.999-82
Telefone1: (41) 3205.8008   Telefone2: (41) 8448.8015
Endereço: RUA NILO PEÇANHA, 86
Curitiba - PR.

TERMO DE AUTORIZAÇÃO DE USO DE IMAGEM E VOZ
Pessoa maior de 18 anos

Neste ato, e para todos os fins em direito admitidos, autorizo expressamente a utilização da minha imagem e voz, em caráter definitivo e gratuito, constante em fotos e filmagens decorrentes da minha participação no Trabalho de Conclusão do Curso TCC de Design de Projeto Visual da Universidade Positivo, a seguir discriminado:

Título do projeto: _Produção Audio Visual: série de vídeos com exercícios para portadores de Parkinson com aporte do design visual_

Pesquisador (es): _Thais de Lima Santos e Thayza Karin de Paiva_

Orientador(a): _Gabrielle A Grimm_

As imagens e a voz poderão ser exibidas: nos relatórios parcial e final do referido projeto, na apresentação áudio-visual do mesmo, em publicações e divulgações acadêmicas e nas redes sociais, em festivais e premiações nacionais e internacionais, assim como disponibilizadas no banco de imagens resultante da pesquisa e na Internet, fazendo-se constar os devidos créditos.

O aluno fica autorizado a executar a edição e montagem das fotos e filmagens, conduzindo as reproduções que entender necessárias, bem como a produzir os respectivos materiais de comunicação, respeitando sempre os fins aqui estipulados.

Por ser esta a expressão de minha vontade, nada terei a reclamar a título de direitos conexos a minha imagem e voz ou qualquer outro.

Curitiba, _11_ de _outubro_ de 2016.

Maria Luiza Mihrick

Assinatura

Nome: _Mª Luiza Mandu Kuiaski_

RG.: _1.720.389.7_   CPF: _402.241.399-91_

Telefone1: (_41_) _3039-7721_   Telefone2: (_41_) _8659-4704_

Endereço: _Rua Raymundo Nina Rodrigues, 261 Capão - Ctba/Pr_

TERMO DE AUTORIZAÇÃO DE USO DE IMAGEM E VOZ
Pessoa maior de 18 anos

Neste ato, e para todos os fins em direito admitidos, autorizo expressamente a utilização da minha imagem e voz, em caráter definitivo e gratuito, constante em fotos e filmagens decorrentes da minha participação no Trabalho de Conclusão do Curso TCC de Design de Projeto Visual da Universidade Positivo, a seguir discriminado:

Título do projeto: *Produção Audiovisual: Série de vídeos com exercícios para portadores de Parkinson com o aporte do Design Visual.*

Pesquisador (es): *Thais de Lima Santos e Thayza Karen de Paiva*

Orientador(a): *Gabrielle Hartmann Grin*

As imagens e a voz poderão ser exibidas: nos relatórios parcial e final do referido projeto, na apresentação áudio-visual do mesmo, em publicações e divulgações acadêmicas e nas redes sociais, em festivais e premiações nacionais e internacionais, assim como disponibilizadas no banco de imagens resultante da pesquisa e na Internet, fazendo-se constar os devidos créditos.

O aluno fica autorizado a executar a edição e montagem das fotos e filmagens, conduzindo as reproduções que entender necessárias, bem como a produzir os respectivos materiais de comunicação, respeitando sempre os fins aqui estipulados.

Por ser esta a expressão de minha vontade, nada terei a reclamar a título de direitos conexos a minha imagem e voz ou qualquer outro.

Curitiba, *11* de *outubro* de 2016.

____
Assinatura

Nome: *Sandra Salomão*
RG.: *1.154.228.0* CPF: *360.739.549-72*
Telefone1: (41) *91372743* Telefone2: (41) *95007921*
Endereço: *R. Bororos. nº 835 - casa Vila Isabel - CEP 80320-260 Ctba - Paraná*

TERMO DE AUTORIZAÇÃO DE USO DE IMAGEM E VOZ
Pessoa maior de 18 anos

Neste ato, e para todos os fins em direito admitidos, autorizo expressamente a utilização da minha imagem e voz, em caráter definitivo e gratuito, constante em fotos e filmagens decorrentes da minha participação no Trabalho de Conclusão do Curso TCC de Design de Projeto Visual da Universidade Positivo, a seguir discriminado:

Título do projeto: _Produção Audiovisual: Série de vídeos com exercícios para portadores de Parkinson com o aporte do Design Visual._

Pesquisador (es): _Thaus Santos e Thayna Karen de Paiva_

Orientador(a): _Gabrielle Hartmann Grim_

As imagens e a voz poderão ser exibidas: nos relatórios parcial e final do referido projeto, na apresentação áudio-visual do mesmo, em publicações e divulgações acadêmicas e nas redes sociais, em festivais e premiações nacionais e internacionais, assim como disponibilizadas no banco de imagens resultante da pesquisa e na Internet, fazendo-se constar os devidos créditos.

O aluno fica autorizado a executar a edição e montagem das fotos e filmagens, conduzindo as reproduções que entender necessárias, bem como a produzir os respectivos materiais de comunicação, respeitando sempre os fins aqui estipulados.

Por ser esta a expressão de minha vontade, nada terei a reclamar a título de direitos conexos a minha imagem e voz ou qualquer outro.

Curitiba, _11_ de _outubro_ de 2016.

Joana Olindina dos Santos
Assinatura

Nome: _Joana Olindina dos Santos_

RG.: _1212 473 2_ CPF: _222 241 069 04_

Telefone1: (41) _9838 1511_ Telefone2: ( ) _______

Endereço: _Inácio Martins 294 casa 9_

Xaxim

TERMO DE AUTORIZAÇÃO DE USO DE IMAGEM E VOZ
Pessoa maior de 18 anos

Neste ato, e para todos os fins em direito admitidos, autorizo expressamente a utilização da minha imagem e voz, em caráter definitivo e gratuito, constante em fotos e filmagens decorrentes da minha participação no Trabalho de Conclusão do Curso TCC de Design de Projeto Visual da Universidade Positivo, a seguir discriminado:

Título do projeto: Produção Audio Visual: série de vídeos com exercícios para portadores de Parkinson com aporte do design Visual.

Pesquisador (es): Thais de Lima Santos e Anayza Karin de Paiva

Orientador(a): Gabrielle H. Grimm

As imagens e a voz poderão ser exibidas: nos relatórios parcial e final do referido projeto, na apresentação áudio-visual do mesmo, em publicações e divulgações acadêmicas e nas redes sociais, em festivais e premiações nacionais e internacionais, assim como disponibilizadas no banco de imagens resultante da pesquisa e na Internet, fazendo-se constar os devidos créditos.

O aluno fica autorizado a executar a edição e montagem das fotos e filmagens, conduzindo as reproduções que entender necessárias, bem como a produzir os respectivos materiais de comunicação, respeitando sempre os fins aqui estipulados.

Por ser esta a expressão de minha vontade, nada terei a reclamar a título de direitos conexos a minha imagem e voz ou qualquer outro.

Curitiba, 11 de outubro de 2016.

____
Assinatura

Nome: Eduardo Dechert filho

RG.: 456711   CPF: 142 127 73953

Telefone1: (   ) 3276805   Telefone2: (   ) 96881674

Endereço: rua Pedro Wobeto 12 Boqueirão

Printed by Books on Demand GmbH, Norderstedt / Germany